ALSO BY ISADORE ROSENFELD, M.D.

Symptoms

Modern Prevention

Second Opinion

The Complete Medical Exam

THE BEST TREATMENT

Isadore Rosenfeld, M.D.

Simon & Schuster
New York London Toronto Sydney Tokyo Singapore

SIMON & SCHUSTER
Simon & Schuster Building
Rockefeller Center
1230 Avenue of the Americas
New York, New York 10020

Designed by Irving Perkins Associates
Manufactured in the United States of America

10 9 8 7 6

Library of Congress Cataloging-in-Publication Data

Rosenfeld, Isadore.
The best treatment / Isadore Rosenfeld.
p. cm.
Includes index.
1. Therapeutics—Popular works. I. Title.
RM122.5.R66 1991
616—dc20 91-30759
CIP

ISBN: 0-671-69339-5

ACKNOWLEDGMENTS

I am grateful to my editor, Fred Hills, for all his help. He and Elaine Glaser, who typed and retyped this manuscript so often I lost count, really deserve an honorary M.D. degree after working with me on so many books. My gratitude also to those colleagues who have taken the time to review the manuscript and who have provided me with so many helpful suggestions:

STANLEY J. BIRNBAUM, M.D.
Professor of Obstetrics and Gynecology
Vice-Chairman, Department of Obstetrics and Gynecology
The New York Hospital-Cornell Medical Center

MYRON I. BUCHMAN, M.D.
Clinical Associate Professor
Department of Gynecology and Obstetrics;
Attending Physician
The New York Hospital-Cornell Medical Center

HARRY L. BUSH, JR., M.D.
Associate Professor of Surgery
The New York Hospital-Cornell Medical Center

JOHN J. CARONNA, M.D.
Professor of Clinical Neurology
The New York Hospital-Cornell Medical Center

MORTON COLEMAN, M.D.
Clinical Professor of Medicine
Attending Physician
The New York Hospital-Cornell Medical Center
Chairman, Fund for Blood and Cancer Research

NANCY H. COLES, M.D.
Assistant in Department of Ophthalmology
Lenox Hill Hospital;
Associate in Department of Ophthalmology
New York Eye and Ear Infirmary

ARMAND F. CORTESE, M.D.
Clinical Associate Professor of Surgery
Associate Attending Surgeon
The New York Hospital-Cornell Medical Center

RICHARD B. DEVEREUX, M.D.
Associate Professor of Medicine
Director, Echocardiography Laboratory
The New York Hospital-Cornell Medical Center

MURRAY DWORETZKY, M.D.
Clinical Professor of Medicine
Attending Physician
The New York Hospital-Cornell Medical Center

PHILIP FELIG, M.D.
Clinical Professor of Medicine
New York Medical College;
Attending Physician, Senior Medical Staff
Lenox Hill Hospital

RICHARD C. GIBBS, M.D.
Clinical Professor of Dermatology
New York University Medical Center

HOWARD GOLDIN, M.D.
Clinical Professor of Medicine
Attending Physician
The New York Hospital-Cornell Medical Center
Consultant, The Hospital of Rockfeller University

WILBUR JAMES GOULD, M.D.
Adjunct Professor of Otolaryngology
New York Medical College;
Clinical Professor of Otolaryngology
New York University;
Director and Founder of Vocal Dynamics Laboratory
Lenox Hill Hospital

CATHERINE HART, M.D.
Associate Professor of Medicine
The New York Hospital-Cornell Medical Center

BARRY HARTMAN, M.D.
Clinical Associate Professor of Medicine
The New York Hospital-Cornell Medical Center

LAWRENCE J. KAGEN, M.D.
Professor of Medicine
Attending Physician
Hospital for Special Surgery
The New York Hospital-Cornell Medical Center

MARILYN G. KARMASON, M.D.
Clinical Associate Professor of Psychiatry
The New York Hospital-Cornell Medical Center

JOHN H. LARAGH, M.D.
Hilda Altschul Master Professor of Medicine
Director of Cardiovascular and Hypertension Center
Chief, Cardiology Division, Department of Medicine
The New York Hospital-Cornell Medical Center

BRUCE B. LERMAN, M.D.
Director of Cardiac Electrophysiology Laboratory;
Associate Professor of Medicine
The New York Hospital-Cornell Medical Center

JAMES P. MCCARRON, JR., M.D.
Clinical Assistant Professor of Surgery (Urology)
The New York Hospital-Cornell Medical Center

MARCUS M. REIDENBERG, M.D.
Professor of Pharmacology and Medicine
The New York Hospital-Cornell Medical Center

HOWARD L. ROSNER, M.D.
Assistant Professor of Anesthesiology
Director Pain Management Service
The New York Hospital-Cornell Medical Center

JAMES P. SMITH, M.D.
Clinical Professor of (Pulmonary) Medicine
The New York Hospital-Cornell Medical Center

DAVID ZAKIM, M.D.
Vincent Astor Distinguished Professor of Medicine:
Director, Division of Digestive Diseases
Attending Physician
The New York Hospital-Cornell Medical Center

For my Camilla—

Again, Still, and Always

No book can replace the services of a trained physician. This book is not intended to encourage treatment of illness, disease, or other medical problems by the layman. Any application of the recommendations set forth in the following pages is at the reader's discretion and sole risk. If you are under a physician's care for any condition, he or she can advise you about information described in this book.

Contents

INTRODUCTION

"Good?" "Better?" "Best?" Is there really any difference between a good treatment, a better one, and the best? How can a patient tell which is which? If one treatment works, how can another be better, or the best? Well, consider this. Suppose you're a fifty-eight-year-old man who has been found to have high blood pressure. Your doctor prescribes a medication and after two weeks your pressure is normal. Unfortunately, however, you now have to get up several times a night in order to empty your bladder because the drug you were given is a diuretic. So that was "good" treatment for the high blood pressure, but it made you miserable in the process. When you tell your doctor about this complication, he says he has something "better" for you, and gives you a beta-blocker. At your next visit, your blood pressure again is excellent. The nightly visits to the john have stopped, too. Everything is just fine *except* that you have now become impotent. That may or may not be "better" than peeing all night, depending on your age and priorities. But you have not yet received the "best" treatment for your hypertension—one that normalizes your blood pressure safely and effectively without these or other side effects. Such medication exists, and is available. But millions of men and women with hypertension, and a host of other disorders, are receiving "good" or "better" treatment, but not the "best."

This book describes in language that everyone can understand what, in my experience as a practicing doctor for over forty years, *is* the best treatment for the many symptoms and diseases you are likely to encounter in your journey through life. I asked several of my colleagues who are distinguished specialists in various fields to review this advice to you, and to add their own recommendations.

The topics discussed are arranged in alphabetical order. My treatment preferences are not necessarily "better" than those of your own doctor, but they may be different. And if they are, then your awareness of these differences may help you, especially if you're not tolerating or responding optimally to the therapy you are now receiving.

For me to go on record this way and endorse a particular approach, from among many, is neither unusual nor presumptuous. Doctors constantly indicate their therapeutic preferences in medical journals and textbooks. That's how we learn from each other. The difference is that they write for their *colleagues*—I write for *you*. Take from these pages whatever you can use, but before acting on any recommendation, *always clear it with your own doctor first*. Don't worry about antagonizing him or her with your doubts, questions, or suggestions. Doctor-patient relationships are not what they used to be! These days the medical profession is well aware of, and often caters to, the public's interest in health care. Most doctors *should* and *will* listen carefully to what you have to say, especially if you're well-informed.

Some of my preferences may surprise you, especially if you haven't kept up with all the breakthroughs in the world of medicine. That's hard enough for doctors to do, let alone patients. For example:

- Aspirin is *not* necessarily the safest way to reduce a fever.
- You may *not* need surgery for an enlarged prostate that gets you up every hour or two during the night.
- You *can* still look forward to having a child of your own even if an exhaustive and expensive fertility work-up suggests otherwise.
- *Not* everyone with high blood pressure need follow a strict low-salt diet.
- A cancer that has spread is *not* necessarily incurable.
- A low cholesterol level is *not* always better than a higher one.
- There *are* more effective ways than calcium supplements to prevent and treat osteoporosis in menopausal women.

There's much that's new in the field of health care—medications to control the pain of arthritis, to manage the premenstrual syndrome, to treat cancer, AIDS, and heart attacks. You'll read about all of them—and more—in this volume.

Once You Know What Your Symptoms Mean

It's usually not a good idea to treat a *symptom* (as opposed to a *disease*) before you know what's causing it. For example, if you have a cough that won't go away, suppressing it with a potent mixture may divert you from finding and curing an underlying pneumonia or cancer. When you hurt somewhere but don't know why, the worst thing you can do is to mask the pain with a pill and then go about your business. You may be ignoring an important early warning signal of a problem that's still curable—but perhaps not for long! The antacid that soothes your "indigestion" may delay the studies that could reveal an ulcer or cancer of the stomach. Temporarily reducing a fever without identifying its source may be equally dangerous. Unless you know *why* you have *any* symptom, and treat its *cause,* today's minor illness can end up as tomorrow's major problem.

But having said all of that, there's no point in continuing to suffer *after a diagnosis has been made.* So, if your dry cough is, in fact, simply the tail end of a common cold, you *should* take an effective cough suppressant; if you've hurt yourself in a fall, you *should* have the best pain killer there is; if you have indigestion due to hyperacidity, you *should* use the most effective antacid available; if you have a high fever from a documented strep throat, you *should* lower it in the safest, most effective way; if an itch is due to poison ivy or to an allergic reaction and is driving you up the wall, you *should* do more than just scratch it; if a "24-hour virus" has left you vomiting, you *should* end the retching *now.*

This book will help you do all that—and more.

ACNE: No Longer a Wallflower at the Prom

The other day I returned for a sentimental visit to my old high school after almost fifty years. Everything was very different from how I remembered it. The building itself seemed smaller, and the students looked so much younger than I did when I was their age! What also impressed me was how many of them had facial acne. I was never really aware of more than an occasional student with "bad skin" in my own class "way back." Was this something new at Baron Byng High? Had it been struck by an epidemic of some kind? Or was it just my faulty memory?

The facts, of course, are that 85 percent of us experience acne sometime during our teens. It is true now and it was true fifty years ago. The disorder is more common and severe in males, and it runs in families, so that if your mother, father, or siblings were or are affected, you're more likely to be, too. Although the incidence peaks at age eighteen, one is vulnerable to acne until thirty, and sometimes beyond that. Given these statistics, I took a hard look at our family album for close-up pictures of myself in my teens. Guess what I found!

There is a great deal of folklore surrounding the causes of acne, most of which revolves around sex and life-style. Much of it is pure nonsense. For example, acne has nothing to do with being oversexed, undersexed, masturbation, or "cleanliness." The role of diet has probably been overemphasized, too. Although I personally don't know of any teenager whose skin erupts from eating too many sweets, fats, or junk food, if you are convinced that something in your diet repeatedly aggravates your condition, then no one can really argue with that observation. However, the real cause of acne is the overproduction of oil (sebum) by glands in the skin in response to stimulation by the male hormone testosterone. Since both sexes produce testosterone (though in different amounts), acne occurs in women as well as in men.

There are new and effective ways to treat acne. The one to choose depends on whether the problem is simply blackheads and whiteheads (which doctors call "open" and "closed" comedones, respectively) or large cysts or nodules that contain pus, rupture from time to time, and often cause scarring and disfiguration. However, regardless of whether your case is mild or severe, keep away from greasy cosmetics. Use only water-based makeup, including hair-styling gels, which can spill onto the skin of the forehead, sometimes a site for acne. Although sun is generally good for this condition, don't get carried away with it. Too much can lead to more serious hazards than acne, especially if you are fair-skinned or redheaded. In addition to avoiding oily cosmetics, you can further reduce excess grease on your face by washing frequently with soap and water. These general measures will usually improve blackheads and whiteheads. If they don't, you'll have to see your doctor.

Most dermatologists begin treatment with *topical* tretinoin (Retin-A)—not to be confused with Accutane, the *oral* form, which can have devastating effects in pregnant women. Topical Retin-A comes in cream, gel, and liquid form (the latter is quite irritating and not routinely used). If your skin is on the dry side, use the cream; if it is oily, apply the gel. But Retin-A in any formulation is somewhat irritating, so always start with the weakest strength (0.01 percent in the gel and 0.025 percent in the cream). Rub a pea-sized amount into the affected area *every third night* about an hour before going to bed after you have washed and dried your face. Don't apply it just as you're getting into bed because the medication may come off on the sheets and pillows before it is absorbed. Whichever form and strength of Retin-A you choose, don't be disappointed if initially your skin becomes red, itches, and peels. That's par for the course. And don't be impatient. This medication usually takes several weeks to work. If you're not clearly improving by the sixth week, increase the strength, but always under your doctor's guidance. Keep out of the sun during the entire treatment period in order to avoid further irritation. When some solar exposure is unavoidable, apply lots of high-grade sun blocker, at least number 15 and preferably higher. If you have big pus-filled pimples, you should also use warm salt water compresses as often as possible—three and four times a day.

I believe you should also have an antibiotic when taking Retin-A. Although this can be topically applied (clindamycin, erythromycin, or benzoyl peroxide), I prefer oral *tetracycline,* 500 mg, twice a day. After you see definite improvement, begin to taper the dose until the acne has cleared completely. But do not take tetracycline if you are under thirteen years of age or are pregnant (because it will stain the teeth of children and those of your newborn). This antibiotic may also sensitize the skin to the sun's rays and can result in a troublesome rash. Some doctors believe too that tetracycline interferes with the contraceptive action of the Pill as well. So if you're looking for a suitable alternative, take oral erythromycin or minocycline (Minocin).

Severe acne may form cysts that later scar and disfigure the skin. If that's what's happening to you, discuss using *oral Retin-A* (Accutane) with your doctor. It will produce dramatic improvement within three to four months. However, I recommend this medication only *as a last resort,* because it can hurt the liver, raise the blood cholesterol, and most important, result in a deformed baby if taken during pregnancy. There were 2,500 pregnant women who used Accutane between 1982 and 1988, and who chose not to have an abortion. Six hundred of them (about 1 in 4) gave birth to infants with serious defects. So I *never* prescribe it to any sexually active woman for whom there is even the *slightest* risk of conception.

If you've been cleared to use Accutane, keep in constant touch with your doctor or dermatologist. And remember, if you are of childbearing age, have a pregnancy test done *before* you start therapy, and monthly for as long as you're taking this medication unless you've had *absolutely no sexual* contact whatsoever. It's also a good idea to monitor your cholesterol and triglyceride levels, which can be substantially increased by this drug. In sunny weather, stay in the shade, cover your body from head to toe, and use a high-grade sun block.

If you have old facial scars from adolescent acne that was inadequately treated, they can often be dramatically improved either by *dermabrasion,* in which the superficial layers of skin are planed down with modern, sophisticated equipment, or by *chemical facial peeling*. These procedures are done by dermatologists and cosmetic surgeons.

AIDS: Cure—Not Yet; Treatment—Sort Of; Prevention—Yes!

AIDS (acquired immunodeficiency syndrome) is caused by a virus that disarms the immune system, leaving the body defenseless against infections that it can normally overcome. So AIDS patients die from what are called "opportunistic" diseases—*Pneumocystis carinii* pneumonia (PCP, which hits the lungs), Kaposi's sarcoma (which affects the skin), and a variety of viral, fungal, and other infections that spread through the gut, the brain, and nervous tissue.

There is no cure for AIDS. The best and only approach to the problem at this time is to learn how to avoid it, and to act on that information. This ultimately fatal infection is spread by: (a) sexual contact with someone of either sex who has the disease, (b) sharing needles for intravenous drug use, and (c) transfusions, which still present a minimal risk despite donor blood screening.

If you are vulnerable to AIDS by virtue of any of these risk factors, and especially if you are a sexually active gay male, never ignore an unexplained cough, fever, or rash, persistent headache, unusual irritability, or other neurological signs (double vision, weakness of an arm or leg, stiff neck). But you can be harboring the AIDS virus and not have any symptoms whatsoever. I have several patients who have continued to test positive for HIV (human immunodeficiency virus)—the virus that leads to AIDS—for as long as ten years, yet who still feel perfectly well and have no evidence of the disease.

What should you do when you have been infected with AIDS but are symptom-free? And what's the best treatment when the disease is in its full-blown state?

When you test positive but don't feel sick, first make sure it's not a laboratory mistake. Such errors do happen. However, if the result is confirmed in a repeat analysis, have a T cell count done every three or four months. Here's why. The T cells in the blood are responsible for mounting an effective immune response. A count *less* than 250 per cc means that your immune system is severely impaired; 250–500 puts you in the gray zone; a reading greater than 500 is reassuring. If you are symptom-free with a count *above* 500, there is nothing you can or should do. A downward trend suggests that you may be approaching the time when the disease will become apparent. At that point I recommend that zidovudine (Retrovir, which used to be called AZT) be started. Retrovir retards the virus's ability to reproduce in T cells. In addition, I usually prescribe trimethoprim-sulfamethoxazole (Bactrim DS, Septra DS), 2 tablets a day, 3 times a week. This combined regimen appears to help delay the onset of the PCP infection in the lung.

Once symptoms have appeared, treatment depends on which system of the body is involved, and by what "opportunistic" disease. For example, if the lung is infected by PCP, you're best off with Bactrim by vein, or with a drug called Pentamidine, also given intravenously. If AIDS has struck the brain with toxoplasma infection, the most effective agents are a combination of pyrimethamine and sulfadiazine. When a fungus is the culprit, the most useful drug is amphotericin B given intravenously. Kaposi's sarcoma, a form of skin cancer that in its late stage can also involve the mouth and the lungs, responds best to recombinant alfa interferon, zidovudine, and some of the anticancer drugs. These medications don't cure AIDS, but do make the patient feel better and slow down the disease process.

In summary, remember this about AIDS: Although it is not now curable, and its treatment is certainly less than satisfactory, its onset and progress can often be retarded. Take advantage of *any* treatment offered to you, always making sure that it is the latest and the best available.

ALCOHOL ABUSE: What Can a "Problem Drinker" Do?

The problems resulting from alcohol abuse are myriad. They include congenital deformities in babies born of mothers who drank even small amounts during pregnancy, as well as broken marriages, wrecked careers, fatal accidents, and death from cirrhosis in adult life. But these are only the highlights of the devastating impact this "pleasant poison" has had on our society. Regardless of whether someone's drinking problem is genetically determined (a theory that has recently been challenged) or due to environmental influences, the inability to cope with "normal" alcohol consumption or the compulsion to drink oneself into oblivion is certainly a disease. And like many other illnesses, it *is* treatable. But the first step toward recovery requires you to be aware of and accept the fact that you have a drinking problem. If you're not sure about it, there are two simple questionnaires that will usually settle the matter. The first, the CAGE test, consists of only four questions.

- Have you ever felt the need to Cut down the quantity of your drinking?
- Do you get Angry when someone tells you that you drink too much?
- Do you feel Guilty after drinking?
- Do you need an Eye-opener or a drink to get you started in the morning?

One yes answer is borderline, but two or more indicate a problem. To confirm it, proceed to the MAST (Michigan Alcoholism Screening Test) questionnaire developed by Drs. Pokorny and Kaplan. Here are the ten questions to be asked:

1.	Do you feel you are a normal drinker?	Y (0)	N (2)
2.	Do friends or relatives think you are a normal drinker?	Y (0)	N (2)
3.	Have you ever attended a meeting of AA?	Y (5)	N (0)
4.	Have you ever lost friends or girlfriends/boyfriends because of drinking?	Y (2)	N (0)
5.	Have you ever gotten into trouble at work because of drinking?	Y (2)	N (0)
6.	Have you ever neglected your obligations, your family, or your work for two or more days in a row because you were drinking?	Y (2)	N (0)
7.	Have you ever had delirium tremens or severe shaking, heard voices, or seen things that weren't there after heavy drinking?	Y (2)	N (0)
8.	Have you ever gone to anyone for help about your drinking?	Y (5)	N (0)
9.	Have you ever been in a hospital because of drinking?	Y (5)	N (0)
10.	Have you ever been arrested for drunk driving or driving after drinking?	Y (2)	N (0)

A score of 6 or more obtained from these ten questions identifies 90 percent of problem drinkers.

Every treatment approach to the problem drinker must include supportive counseling to provide motivation and the realization that there *is* a problem and that it *can* be solved.

With regard to specific therapy, the two main objectives are: (1) Cessation of drinking and relief of withdrawal symptoms, and (2) Ongoing efforts to ensure permanent abstinence.

There is no medication that will "cure" an alcohol problem. Some doctors prescribe disulfiram (Antabuse), a chemical that, when mixed with alcohol, produces so violent a reaction that even its anticipation terrifies the drinker. On the recommended dose of 250 mg a day, just an ounce or two of hard liquor, a can of beer, or just one glass of wine will produce symptoms—severe headache, nausea, vomiting, flushing, and rapid heart rate—that you will not soon forget! Ask anyone to whom it's happened. But Antabuse is not an effective deterrent, because "once bitten, twice shy." Anyone who *wants* to drink is free simply not to take Antabuse a second time. To overcome this drawback, a method has been devised to implant Antabuse under the skin so that, like it or not, it's with you always (until surgically removed). Now, just you dare to have even one drink! Although it sounds like a solution, there are problems with this therapy, too. Compulsive addicts who *must* have their drink may then suffer so violent a reaction as to become dangerously ill if they happen to have some other serious underlying disease (heart trouble, a neurological disorder, and so on). In any event, this new delivery system, which is still being investigated, may work for selected problem drinkers.

If you've decided to pass up the Antabuse and to go "cold turkey" on your own, the warning signs of impending DT's (delirium tremens) are restlessness, anxiety, panic, insomnia, and nightmares. In the full-blown attack, there is also confusion, agitation, and hallucination, symptoms that occur when *any* narcotic or "recreational" substance is suddenly withheld. At this point, you belong in the hospital where fluids, vitamin supplements (especially thiamin and the rest of the B group), and sedation with one of the benzodiazepine drugs (Librium, Valium, Ativan) can be administered.

A better approach, in my view, is a "detoxification" program in which alcohol is *gradually* replaced by "cross tolerant" and "cross dependent" drugs. This requires your going to a clinic where diazepam (Valium) is administered around the clock in order to avoid the DTs. You'll need a minimum of three days to get the booze out of your system, but in order to *stay* "dry" over the long term, I advise you to "live in" for about a month, if at all possible. On the average, only one in three patients remains on the wagon for any length of time after detoxification.

There are several such facilities throughout the United States and Europe, the prototype of which is the Betty Ford Center in Palm Springs, California. If you decide to try this treatment, find out all you can about the clinic you're planning to attend. The best way to do that is to ask someone who has been there.

The program my "problem drinker" patients have found most effective over the years is also the most famous one—*Alcoholics Anonymous*—a truly wonderful nonprofit organization. The loyalty it generates is astounding! At the start, you attend meetings daily, sometimes even several times a day if necessary. You are encouraged to share your problems with others, and soon feel comfortable doing so. But AA is not just a bunch of drinkers chewing the fat; it's a psychologically sound support system that I encourage anyone with an alcohol problem to try. In fact, most detoxification centers advise their patients to follow up at AA after they have dried out. Al-Anon, a parallel organization whose goal is to help relatives of problem drinkers, is also worth looking into if someone you care about is in that kind of trouble.

In summary, although problem drinking is a disease with many complicated causes, both genetic and environmental, it can respond to pharmacological treatment and to psychological support. The latter, in my experience, is best obtained over the long-term through Alcoholics Anonymous.

ALLERGIES: When Your Nose Runs and Your Eyes Itch

Allergies have many targets. Some of them are discussed elsewhere in these pages—the skin (with its hives and rashes), the gut (diarrhea), and the air passages (asthmatic wheezing). This particular section is all about the nose. The allergic nose, provoked by something you inhale, leaves you with a variety of symptoms that doctors call *allergic rhinitis*. For example, my wife loves all cats, especially fluffy Angora kittens and the sleek Siamese breeds. But the moment she comes near one, her eyes and nose begin to run and itch. Unless she and the animal quickly part company, she becomes really uncomfortable. Ironically, I, who have no interest in or particular love for cats, experience absolutely no adverse reaction to them no matter how many there are or how close the contact.

About 40 million Americans (more than 10 percent of the population) develop allergic rhinitis in response to a variety of inhaled substances. In my wife's case, it's cat dander; one of my sons can't tolerate dog dander; for others it's pollen, or house dust (in which the offending agent is the ubiquitous microscopic mite), molds, a host of commercial inhalants, wood dust, and certain enzymes present in household detergents. You can occasionally spot someone with allergic rhinitis by the dark circles under their eyes, aptly called "the allergic shiner." This discoloration is probably due to chronic nasal obstruction, from any cause and so also occurs in those whose stuffiness is nonallergic in origin.

The best way to deal with allergic rhinitis is, of course, to avoid or at least to minimize your exposure to whatever offends your nose. For example, if you're allergic to mold and mite dust, reduce the humidity in your home (both thrive in a damp environment), vacuum regularly, and clean your draperies and carpets every few months; find a new home for the pet that's making you sick; demand a tobacco-free work place; stop using cosmetics and polishes whose fumes make you sick.

However, in the real world there are, unfortunately, some inhalants from which one just can't escape. It's all very well for me to advise you to leave town for several months during the pollen season, or to relocate permanently, say, to Arizona, but that's not feasible for most people. So at some point you're usually going to have to treat the allergic symptoms themselves. Here are the best ways to do so.

Antihistamines provide the most effective relief for allergic rhinitis. They neutralize "histamine," the chemical culprit responsible for all the symptoms. Although virtually any antihistamine will help, the major drawback to most brands is the drowsiness they produce in more than half the people who take them. A sedative effect is welcome at bedtime, but not during the day. So the antihistamines I recommend are terfenadine (Seldane)—daily dose is 2 tablets, 60 mg each—or astemizole (Hismanal)—one 10 mg tablet daily—because they are much less likely to make you drowsy than most of the others. Remember, however, if you have an enlarged prostate, *any* antihistamine can interfere with normal voiding of urine—a potential emergency situation.

Don't use nose drops for more than a few days when your nose is "blocked" and/or running, because after the dribbling stops, your nose will feel stuffier than ever.

If the antihistamines don't work, cortisone-type sprays will. In its topical form, this hormone is only minimally absorbed. Whatever small amount you swallow is inactivated as it passes through the liver. The various preparations on the market—Beconase, Vancenase (the same product with different trade names), and Nasalide—are equally good. Ask your doctor to prescribe the least expensive one. Sprayed into the nose twice daily for two weeks, this medication will almost always eliminate for several weeks the nasal congestion, itching, sneezing, and runny nose due to chronic exposure to some inhalant you simply cannot avoid.

Should you be desensitized to your allergen? Desensitization involves receiving tiny doses of the offending substance at regular intervals until your body builds up "immunity" to it. I do not routinely recommend desensitization because it takes years to complete, it's costly, and it can sometimes produce unpleasant side effects. But if your symptoms persist, or if you do not respond to treatment and there's no way of avoiding the allergic stimulus, then desensitization is worth doing.

ALZHEIMER'S DISEASE: More Than a Matter of Aging

When I was a medical student, the designation "Alzheimer's" was reserved for persons in their forties, fifties, and sixties with disabling and progressive deterioration of memory. Similar complaints in the more elderly were taken for granted as the inevitable and expected consequences of growing old. In other words, memory loss was considered to be "abnormal" in the young, but merely a sign of "normal" aging in older individuals. This distinction was all wrong.

Forgetting a name occasionally, or where you put your keys or even where you parked your car, is no cause for alarm. When a younger individual has a "bad memory" and can't always come up with a name on short notice, it's often the result of anxiety, inattention, or distraction. So if you're young and want to be sure not to forget something important, make notes. Real Alzheimer's is not just an occasional lapse or embarrassment; it's a gradually worsening impairment of memory that ends in the total inability to function. What's more, *it's a disease at any age.*

The typical case of Alzheimer's begins with predictable and *repeated* loss of memory for recent events (such persons remember, at least for a while, what happened fifty years ago but not what they ate for breakfast ten minutes earlier), and progresses to obvious confusion. In the late stages of the disease, the patient can no longer care for himself; he or she cannot be trusted with medication, is unable to prepare meals, attend to basic toilet requirements, or dress. As time goes by, this decline is accompanied by disturbing emotional and personality changes. Eventually, these unfortunate souls cannot even communicate; they lose most of their coordination and are finally chair-bound or bedridden.

This devastating disorder constitutes an enormous emotional and physical burden on other family members, and when admission to a nursing home or some other such facility becomes necessary, as it often does, sadness in the family is often compounded by guilt.

While there is no specific treatment that will restore memory or allow patients with Alzheimer's to function normally, there are drugs that can keep them comfortable. Much of the care given in some nursing homes consists of oversedation, which hastens death. In fact, more than half of all nursing home residents, no matter why they're there, usually receive two or more tranquilizers or other drugs that act on the brain.

It is extremely important *to make sure that the diagnosis of Alzheimer's is correct*. Since there is presently no specific test to confirm it, other causes of physical and behavioral deterioration should always be ruled out. In one autopsy series some years ago, many of those diagnosed with Alzheimer's before death were found, too late, to have had the following reasons for their symptoms:

- nutritional deficiencies (especially of the B vitamins);
- overmedication with sleeping pills, tranquilizers, anti-anxiety preparations, and other psychotropic agents;
- multiple small strokes (usually due to poorly controlled high blood pressure);
- depression (loneliness, bereavement, or other adversities);
- low thyroid function;
- AIDS (a newcomer on the scene);
- brain tumors;
- subdural hematoma (the result of even a minor blow to the head causing blood to accumulate under the skull, press on the brain, and result in its malfunction);

- unrecognized infection (syphilis was formerly a common cause of such behavioral changes and should not be overlooked even today. However, the current focus is on Lyme disease. Not infrequently, the brain of someone thought to have died from Alzheimer's is found to contain the Lyme spirochete).

So before you conclude that an older relative or friend who's behaving "strangely" has Alzheimer's, ask their doctor whether these other disorders have been considered and excluded.

When the brain of a patient with Alzheimer's is examined after death, characteristic changes are apparent under the microscope. Unfortunately, we have no idea what causes them or what they mean. For example, unusually large amounts of aluminum may be present in the tissues. How it got there and why remains a puzzle. This observation has led to the theory that aluminum is in some way responsible for the symptoms of Alzheimer's. I know many sophisticated individuals, including some scientists and doctors, who avoid aluminum in all its forms, from pots and pans used in cooking to those oral antacids and underarm deodorants that contain aluminum. But although the suspicion exists, the proof does not.

There is one observation with which everyone agrees: The brain in Alzheimer's is deficient in an enzyme required for the synthesis of a chemical called *acetylcholine*, which is responsible for the transmission of nerve "messages" or impulses. Most of the current research and "treatment" in this disease is directed at finding and administering agents that increase the amount of this missing acetylcholine. Frankly, I have not found any of the products now available to be effective, although some doctors are convinced that they are of some help. Many prescribe ergoloid (Hydergine) mainly, I think, because it leaves them and the family with the feeling they're doing "something." The recommended dosage is 1 to 2 mg, 3 times a day, but some doctors believe that larger amounts, up to 9 mg a day, improve behavior and/or memory slightly. I'm skeptical because I have never seen any significant response to Hydergine in any of my own patients. In fact, I have stopped prescribing it since reading a recent report suggesting that it may, in fact, have *adverse* long-term effects in some persons.

Lecithin, a complex body fat present in high concentrations in the brain and nerves, and *choline,* a constituent of Vitamin B complex, are both precursors of acetylcholine (a "precursor" is a building block in the manufacture by the body of a given substance) and are also widely given to persons with Alzheimer's. Their main food source is egg yolk, but most people buy the supplements in health food stores. Neither has been shown to have any effect whatsoever in most objective tests.

Physostigmine (*Antilirium*) neutralizes an enzyme that reduces the amount of acetylcholine in the brain. There have been some reports that when taken for several months, this drug may result in very slight improvement in memory. Again, I'm not convinced.

You may have read about THA (tetrahydroaminoacridine), a brain stimulant similar to physostigmine, which was widely reported in the media to improve the symptoms of Alzheimer's disease. As a result of these initial enthusiastic although largely anecdotal reports, the FDA undertook to evaluate it scientifically; the results to date have revealed conflicting conclusions. Unlike Hydergine, lecithin, and choline, THA can temporarily hurt the liver until the dosage is reduced or the drug is discontinued. Nevertheless, the FDA has decided to make this agent available for those who wish to try it.

If a person has had multiple strokes and manifests Alzheimer's-like behavior, an aspirin a day may prevent further deterioration and even result in some improvement by keeping the blood flow going within the brain. When someone with Alzheimer's is agitated, fearful, belligerent, or delusional (they are rarely violent), there are several calming agents that can help. But these drugs should be used *only* when needed and not simply because the individual is old, a "nuisance," and/or in a nursing home. Small amounts of propranolol (Inderal), 10 mg, 3 times a day, have a calming effect in irritable and restless individuals. When there is great agitation, I prescribe lorazepam (Ativan), 1 mg, 3 times a day. In cases of depression, I prefer imipramine (Tofranil) in a dosage of 20 to 40 mg a day.

It has been estimated that because of our aging population, there will be five times as many cases of Alzheimer's within the next fifty years as there are today. Hopefully, by that time we will understand more about what causes this disorder, and how to prevent and treat it. In the meantime, love and compassionate care, either at home or in an institution, together with drugs to control troublesome symptoms, are the best we can do.

AMOEBAE: When You've Ignored the Advice Not to Drink the Water

Here's a scenario that may be familiar to you. You've just returned from a trip to Latin America, Asia, Africa, or indeed anywhere. You think you've picked up an intestinal infection from some food or water along the way because you've come home with diarrhea, cramps, abdominal pain, blood in the stool, and maybe even a little fever. You're not sure what to do about it. You've heard about "turista" and you know that it sometimes clears up just with home remedies. You're not keen on seeing a doctor because you're afraid that he's going to do things to you that you won't particularly enjoy. So you take some of the antidiarrheal preparation left over from the last trip—Imodium, Lomotil, Pepto-Bismol, or one of the others. You may feel better for a day or two, but then the diarrhea starts up again. At this point, you throw in the towel and visit your doctor. He doesn't do what you feared he would, but instead simply asks for a fresh stool specimen. The report comes back that it's teeming with amoebae, a single-cell parasite invisible to the naked eye. This was not the simple "turista" you expected.

Almost 500 million people are infected with amoebae worldwide, and thousands die from it every year. The organism is usually acquired from water, food, or personal contact with a carrier. The most common symptom it causes is dysentery. If they are not eradicated from the bowel, amoebae can invade other parts of the body such as the liver or even the lung.

The drug of choice for amoebiasis is Flagyl (*metronidazole*) normally taken by mouth (if you're very sick, your doctor may give it to you intravenously). Flagyl comes in 250 mg doses, and 3 a day for 10 days will almost always cure the infection. This treatment may leave you nauseated with a metallic taste in your mouth, but usually not severe enough to warrant stopping the drug. *You must abstain from any form of alcohol while you're taking Flagyl.* If you don't, you'll suffer the same side effects as do problem drinkers who try to sneak one in while on disulfiram (Antabuse)—severe belly ache, nausea, and vomiting. Flagyl can also result in fetal abnormalities, so do not use it when you're pregnant.

Sometimes after a complete course of therapy with Flagyl, amoebae remain in your stool even though the diarrhea has cleared. Congratulations!

You are now a carrier! In that case, Yodoxin (iodoquinol), 650 mg, 3 times a day for 20 days, will usually effect a cure. But don't take the Yodoxin for any longer than that because it can then damage the optic nerve and lead to blindness.

There is a new chapter being written in the amoebiasis story. It is estimated that amoebae eggs are present in the stools of almost 40 percent of homosexuals in this country alone. And most of them are what is called silent carriers, without any symptoms whatsoever. Some doctors feel these individuals should be left alone as long as they continue to feel perfectly well. However, in my opinion, they should be treated—for the following reasons: Homosexuals are generally at increased risk for AIDS, and there is some evidence that the presence of amoebae anywhere in your body speeds the rate at which AIDS progresses. Also, we are seeing more and more serious complications like amoebic liver infections and abscesses in infected persons who have no symptoms. So it's best to eradicate the organism even if it's "silent."

The most effective drug for the treatment of asymptomatic amoebiasis is paromomycin sulfate (Humatin). Take 25 mg for every 2 pounds of body weight per day for as long as the bug continues to show up in your stool. (If you weigh 200 pounds, that would be about 750 mg, 3 times a day.) If you are a sexually active homosexual, make sure that your partners are also treated; otherwise you'll simply keep reinfecting each other.

ANEMIA: Energizing Tired Blood

The advertising media have made "anemia," "iron deficiency," and "tired blood" household terms. As a result, many people wrongly believe that anemia is *always* due to iron deficiency, and that taking extra iron is all you need to do to cure it. The fact is, you can be anemic even when your body is loaded with iron. For example, if your family roots are in the Mediterranean basin—Italian, Greek, or Arabic—you may have thalassemia, a type of anemia in which iron stores are plentiful. Also, certain cancers, liver and kidney disease, and chronic infection may all produce anemia despite lots of available iron. So if you're anemic, make sure it's the iron-deficiency type before you rush out to buy an iron supplement for your "tired blood."

Most normal adult men, and women who have stopped menstruating, do not need and should not take supplemental iron. In fact, doing so may mask an underlying serious disease. But if you fall into any one of the following categories, you are vulnerable to iron deficiency anemia.

- *Infants and children,* who need extra amounts of iron because they're growing fast and who, for some reason, aren't getting it (too much junk food, families too poor to provide the necessary diet).
- *Pregnant women,* who must provide enough iron not only for themselves, but for the fetus as well.
- *Menstruating females,* who lose significant amounts of blood (and therefore iron) every month.
- *Long-distance runners,* who may lose blood from the circulation while running.
- *The elderly,* whose diet is inadequate for a variety of reasons—bad teeth (so that they can't chew the necessary iron-rich meats), poverty, or poor absorption from the bowel. But if you're older and anemic, be very careful about *assuming* that it's due to poor nutrition.

Other important causes of "tired blood" in any age group are subtle blood loss (like that from chronic hemorrhoidal bleeding or unrecognized oozing of blood from the stomach in those regularly taking aspirin or any of its newer substitutes, NSAIDs—non-steroidal anti-inflammatory drugs), chronic infection, or even a hidden cancer. Taking iron supplements before these and other causes for the anemia have been found and corrected may mask a life-threatening condition and delay treatment in its early curable stage.

If you are truly iron-deficient, replenish this mineral with food, not pills. It may take a little longer, but it's better in the long run. Of course, if you've had a massive hemorrhage and are *very* iron-depleted, you're going to need immediate replacement, first with a blood transfusion and then with oral iron. But in most other circumstances, an iron-rich diet is all you require—and the best source is red meat (especially organs like liver). Fish, eggs, chicken, dried fruits, enriched and whole grain cereals, bread, pasta, nuts, and dried beans also provide iron, but not as much as does meat. Popeye's claim notwithstanding, spinach is *not* the best or even a particularly good source of iron because the body does not absorb iron efficiently from the leafy greens. In fact, some researchers believe that large amounts may actually impair the immune system. Popeye's biggest mistake, however, was believing that he needed *any* additional iron at all, given the fact that he was a male—men rarely need extra iron.

Here are some additional ways in which you can maximize the iron intake from your diet. Never have tea or coffee *with* your meals because they interfere with the absorption of dietary iron by the gut. If you're anemic from iron deficiency, some extra vitamin C may increase the absorption of the iron in your food.

If you must supplement your iron intake, there are scores of commercially available preparations from which to choose. Most contain basically three types of iron—*ferrous sulfate* (the most frequently prescribed), *ferrous gluconate,* and *ferrous fumarate.* They come in tablets, capsules, and liquids. Children given the liquid should sip it through a straw because iron stains the teeth brown. Make sure, too, that they brush immediately afterward and drink lots of water with every dose. In adults, I have found ferrous fumarate best tolerated. The usual dose is 1 tablet (containing 60 mg of elemental iron) 3 times a day. Try it on an empty stomach first for maximal absorption, but if you end up with diarrhea, constipation, or cramps, then take it with your meal along with 500 mg of vitamin C. (Ferancee has the vitamin C built into the tablet.) Enteric coated tablets cost more, but they're better tolerated, and in the end are probably worth the extra expense. I do not frequently recommend the gradual-release form, which, incidentally, should always be swallowed whole and never crushed or chewed, because these preparations are more expensive and sometimes pass through the intestinal tract intact, without having been absorbed at all!

Virtually any brand of iron can irritate your stomach and give you "indigestion." And don't be alarmed when your stool turns dark green or black; that's just the action of the iron. Too much iron taken indefinitely can be dangerous, so don't take any more than you need. Extra iron that the body can't use is deposited in the liver where it can cause diabetes, impotence, and certain forms of cancer. After your anemia has been corrected and you've stopped the iron supplements, be sure to maintain a diet that provides ample amounts of this necessary mineral—in short, lots of meat and fish.

There are certain substances that *block* the absorption of iron from the gut, no matter how much is present in your diet—medications like Questran (a cholesterol-lowering drug), vitamin E, any antacid, the antibiotic tetracycline, tea, and coffee. So if you are taking any of these, be sure to have them at least 3 or 4 hours *after* you've consumed your iron, either in food or in supplements.

Iron by injection is especially popular in Europe, where patients apparently prefer one injection a month to 3 pills a day. But don't *you* fall into that trap! Take your iron supplements orally unless you have intestinal problems that preclude it, like an acute ulcer, ulcerative colitis, or Crohn's disease, or if large portions of your gut have been surgically removed so that the oral iron cannot be absorbed. I shun injectable iron because it is usually painful, it stains the skin (except when given by vein), and there's at least the suspicion that repeated shots can cause cancer of the muscle into which the iron is introduced. Occasionally, too, you're left with a hard calcium plaque where the injection was administered.

After a hemorrhage or a major operation, when your anemia is profound, you may need *blood transfusions.* Although generally safe, these do carry small but potential hazards *except* if you're receiving your own previously stored supply. The greatest fear people have when given "anonymous" blood is contracting AIDS, but that risk is really quite low these days, since all blood is screened for AIDS antibodies. Of course, it is possible that the blood you received was donated by someone harboring the virus before antibodies developed. Without such antibodies, the blood would test "AIDS-free." In other words, there is a "silent" period in which donor blood *appears* to be healthy due to the absence of antibodies even though the actual AIDS virus itself is there. Statistically, the chances of that happening are about 1 in about 40,000 or 50,000.

Hepatitis, the other danger in transfusion, used to be more important than it is today. Since donor blood can now be effectively screened against the main forms of hepatitis—A, B, and C—that risk is quite small. Still, blood transfusions should only be given when absolutely necessary.

In summary, not every case of anemia is due to "tired blood" or iron deficiency. When iron supplements are necessary, try food first (plenty of meat) and forget about Popeye and his spinach. If you need pills, try ferrous fumarate first, between meals. If that causes side effects, have it *with* your meals, but then don't expect as rapid a replacement because the absorption of iron when mixed with food in the stomach is reduced by more than half. Vitamin C added to the iron usually increases its absorption. Remember that iron should not be taken in unlimited amounts. The excess over that which is actually needed is deposited in vital organs where it can result in serious complications. I do not normally favor or prescribe iron by injection. It's uncomfortable and not without side effects. Discuss with your doctor *before* an operation what his plans and philosophy are with regard to blood transfusions. They can be life-saving when really needed, but carry with them a slight risk that should be avoided if possible.

ANGINA: Drugs, Bypass Surgery, "Ballooning," or Lasers?

Angina pectoris—the pain, pressure, heaviness, or discomfort in the chest that is usually brought on by effort or emotion—is the "cry" of a heart that is deprived of oxygen. This almost always happens as a result of obstruction or narrowing of one or more coronary arteries, which deliver blood to the cardiac muscle. The process responsible for this arterial blockage is called *arteriosclerosis* (hardening of the arteries). Plaques made up of cholesterol, other fats, certain blood constituents, and calcium are deposited on the inside of these vessels, progressively reducing the channel through which the critical blood supply flows. When the disease reaches the stage at which there is too little blood being delivered to satisfy the normal oxygen needs of the heart muscle, angina ensues. Complete closure of a coronary artery, which is often but not always preceded by angina, usually results in a heart attack. One can never predict with certainty when a heart attack will happen; you may have plenty of warning—days, weeks, months, or years after the onset of angina—or it may occur suddenly "out of the blue"! My father first experienced angina at age forty-seven, and did not have a heart attack until he was eighty-three years old! His narrowed arteries did not completely close until forty-five years after his first episode of chest pain, and he remained active all that time despite his symptoms.

The prime objective in treating angina is to control its frequency and severity so as to permit a near normal life-style. The second goal is to delay or prevent a heart attack. Both are possible, but *every treatment regimen must be individually tailored.* The "best" therapy requires the right combination of drugs and changes in life-style when necessary, so that cigarette smoking is terminated, overweight is corrected, a supervised exercise program begun, elevated blood pressure normalized, a diet low in saturated fats and cholesterol carefully followed, and medication to optimize the blood fat picture when diet alone is not successful. No "heart pill" is going to make a real difference to your angina or your long-term survival if you remain fat and physically lazy, continue to smoke, pay no attention to your high blood pressure, and eat the kind of foods that perpetuate high cholesterol levels in your arteries.

There has been a veritable explosion in the past few years in the availability of new and effective drugs to prevent and treat heart pain. These agents, either alone or in combination, make it possible for most persons with angina to lead a virtually normal life. In addition to medication, such patients can now also have their clogged coronary arteries opened mechanically by balloon angioplasty, unclogged by laser treatment, or "bypassed" surgically. In *balloon angioplasty,* a small wire with a deflated balloon at its tip is threaded into the artery and positioned beside the plaque that is narrowing it. The balloon is then inflated, compressing the plaque against the wall of the artery and widening the channel through which the blood flows. *Laser therapy,* a newer and still experimental procedure, "zaps" or vaporizes the obstructing plaque with a laser beam. *Bypass surgery* (now being done more often in this country than the appendectomy) introduces a substitute artery or vein from elsewhere in the body to do the job of the obstructed coronary vessel.

I rarely recommend any of these "invasive" options as long as the angina is stable. I believe that medication should always be tried first. But when less and less exertion produces more and more chest pain, or when the angina occurs at rest or awakens you during the night, it's time for a more aggressive approach. That means an angiogram, in which dye is injected into the coronary circulation. The entire procedure is filmed, permitting doctors to see the arteries and their branches and to decide how best to improve the circulation within the heart.

Drugs that prevent or relieve angina are not pain killers; they work by actually increasing the available oxygen supply to the heart muscle. Some medications do that by dilating the coronary arteries allowing more blood to flow through them; others slow the cardiac rate and lower blood pressure, thus reducing the heart's work load and its energy requirements. Often, several different drugs are prescribed in order to attain both these objectives, and as the condition progresses, doses may need to be increased or medications changed.

Patients often ask how angina can possibly improve or disappear without angioplasty or surgery, given the fact that the disease that narrows the coronary arteries is progressive. The answer lies in the "collateral circulation," a network of microscopic channels within the heart muscle too small to be seen by the naked eye. As the *visible* coronary arteries narrow, these collaterals progressively dilate, and ultimately provide the necessary blood flow—nature's very own bypass!

Medication not only controls symptoms, it also buys time for collaterals to develop. When these do not dilate as quickly as the larger arteries are closing, bypass or balloon or laser therapy becomes necessary.

Here are some of the drugs you should be taking if you have angina. The gold standard is *nitroglycerin*—the little white pill you slip under the tongue either to *prevent* symptoms (when you know that something you are about to do will surely produce them—walking uphill after a heavy meal, especially in cold weather, intercourse, dancing), or to *relieve* them. *No one with chronic angina should ever be without nitroglycerin on their person.* This drug works by widening the coronary arteries and lowering the blood pressure.

Nitroglycerin comes in various strengths and preparations. I normally first prescribe a 0.4 mg dose, and if that is not effective, I increase it to 0.6 mg. The tablet form loses its effectiveness after six months, so get a fresh supply twice a year even if you have lots left in the bottle. Also, since it is inactivated by light and heat, carry only a few in your pocket or purse and store the rest in the refrigerator in the original brown bottle.

I actually prefer the nitroglycerin that comes in a plastic spray bottle (Nitrolingual spray). One squirt under the tongue delivers an exact, metered amount of the drug, which takes effect immediately. The spray has many advantages over the tablet: It is not degraded by light, does not crumble into powder, and you don't have to refrigerate it or discard unused quantities. There is yet another plus. If you are out in the cold and develop angina, instead of groping in your pocket for the little bottle, unscrewing the cap, retrieving one tiny pill (not so easy to do when you're wearing gloves or mittens), and slipping it into your mouth, you simply reach for the plastic container of Nitrolingual spray and squeeze a dose under the tongue.

Sublingual nitroglycerin, by tablet or spray, relieves angina within a minute or two. If it takes ten or fifteen minutes or longer to do so, the "chest pain" for which you took it was probably due to something other than angina. Although it begins to work very quickly, nitroglycerin's action is of short duration, usually five to fifteen minutes. Wait for at least five minutes before taking a second dose. If the chest pain or pressure either persists or, having cleared, recurs moments later, you may be having a heart attack, especially if you've broken out into a cold sweat. If in doubt, call your doctor immediately.

If your current nitroglycerin supply is still fresh, you will almost certainly experience a brief headache after taking it. That's not a side effect, but evidence that the drug is working, for as it dilates the coronary arteries in the heart, it is also widening the blood vessels in the head, and that's what gives the headache. Relaxation of the larger arteries in the rest of the body may cause the blood pressure to fall, too. When the drop is abrupt and substantial, you may faint. Since that's more likely to happen in the standing position, you should always sit down either just before or soon after taking nitroglycerin.

I once found myself in a situation where a potential nitroglycerin faint might have had disastrous consequences for me. Some years ago I was called as a consulting physician to see an affluent European businessman, a cardiac patient. He owned a helicopter, which he dispatched to the airport to bring me to his home in the suburbs. Shortly after the chopper took off, the pilot, who spoke no English and whose native language I did not understand, took a bottle of pills from his pocket, and with his gestures asked whether I approved of them. The container, to my horror, was labeled "nitroglycerin"! There we were at two thousand feet, just he and I, and he was asking whether his nitroglycerin was "okay"! The next thirty-five minutes were among the most harrowing in my life. What would I do if he were to develop chest pain at the controls of the helicopter—encourage him to take the nitroglycerin and run the risk of his passing out, or have him endure the pain and possibly suffer a heart attack?

Thankfully, I never had to make that decision. But the fatigue from my overnight flight compounded by the stress to which I had just been subjected on the helicopter had taken their toll because my patient seemed very concerned when he laid eyes on me! After I recounted the nitroglycerin "adventure," he exploded with laughter. It turns out he had sent the pilot into town to buy the nitroglycerin tablets for him, and the chap was simply checking with me to see whether he'd gotten the right medicine for his boss!

There is also a long-acting nitroglycerin, isosorbide dinitrate (Isordil, Sorbitrate, Dilatrate, and others), which, when taken 3 or 4 times a day, helps *prevent* angina. It comes in a variety of strengths ranging from 2½ mg to 40 mg. The optimal dosage in any given individual can only be determined by trial and error. I usually prescribe the 10 mg dosage, 3 times a day before meals, to start, increasing the strength gradually to the 40 mg preparation. Unlike the sublingual form, isosorbide dinitrate does not have a limited shelf life and need not be refrigerated. Like the sublingual tablet or spray, it, too, lowers blood pressure—but does so gradually, not abruptly. So if you have angina *and* a low pressure, this may not be the right medication for you. Remember, too, that a headache from isosorbide as from nitroglycerin is evidence of effectiveness, not toxicity. Such headaches are usually temporary, they decrease in severity, and finally disappear as you continue to use the medication. Until they do, however, you can obtain relief with aspirin or Tylenol. But try to stay with the nitrates if you possibly can. On the whole, they have fewer side effects than most other anti-anginal agents.

Nitroglycerin is one of an increasing number of drugs now being delivered transdermally—that is, by absorption from the skin. When first introduced, the topical creams and ointments were messy, but they worked, and their effect lasted several hours (as compared to the five to fifteen minutes of the sublingual route). Transdermal nitroglycerin is now available in various strengths as a "patch" (Transderm-Nitro, Nitro-Dur, Minitran, Deponit), which you apply every evening or morning anywhere on the body *above the belly button.* (It doesn't seem to work as well when placed on the legs, buttocks, or lower abdomen.) Remove it after twelve hours. If the patch is left in place for the full twenty-four hours, tolerance to the nitroglycerin may develop. If the plaster on the preparation you're using irritates the skin, switch brands.

When long-acting nitroglycerin preparations do not prevent your angina, you may need to add a *beta-blocker,* which does so by slowing the heart rate and reducing the blood pressure. There are several preparations on the market, each somewhat different from the others. They currently include propranolol (Inderal), metoprolol (Lopressor), atenolol (Tenormin), pindolol (Visken), acebutolol (Sectral), timolol (Blocadren), and nadolol (Corgard), and new ones are being approved every year. I do not usually give beta-blockers to patients whose heart muscle is "weak" (as manifested by shortness of breath and swollen feet). Also, if your cardiac rate is less than 50 per minute, or if you have asthma, skip the beta-blockers and use one of the alternative medicines described below.

Beta-blockers (especially when used together with nitrates) often dramatically improve angina, but can have unpleasant side effects. For example, they may cause profound fatigue; men frequently lose their desire and capability for sex; bad dreams are not uncommon; if you have a blockage of the leg arteries, the cramps in your calves when you walk may be worsened; your memory may become less sharp. Mind you, most people do tolerate these drugs, but if you don't, and reducing the dose doesn't help, you should take something else. Don't just grin and bear it—even if it does help your angina. There are alternatives!

Calcium entry blockers (also called calcium channel blockers) are yet another family of anti-anginal drugs. The one that I have found most effective against angina among the several available is nifedipine (Procardia, Adalat). Its major actions are prevention of spasm of the coronary arteries (an important component of angina in some patients) and reduction of blood pressure. Start with 30 mg a day, taken as 1 slow-release pill (Procardia XL) in the morning, and increase it if necessary to a maximum of 120 mg a day. Its main side effects are flushing of the face, headache, excessive drop in blood pressure, and swelling of the feet. These complications are not usually troublesome enough to warrant stopping the drug if it controls your angina. A newer version of nifedipine, nicardipine (Cardene), is used in much the same way and seems to have fewer side effects in some persons. It comes in 20 and 30 mg strengths and should be tried if the nifedipine is not well tolerated.

Diltiazem (Cardizem) also improves angina, prevents arterial spasm, and lowers blood pressure. The usual starting dose is 30 mg, 3 or 4 times a day, increased if necessary to a total of 360 mg a day. Cardizem comes in a slow-release form, too, so that you need generally take only 1 or 2 tablets a day depending on how much you require. Among the various calcium channel blockers, this one is probably the best tolerated.

Verapamil (Calan, Isoptin), another calcium channel blocker, helps control angina, but I prescribe it more often to lower blood pressure. Your doctor is not likely to recommend this drug unless your heart muscle is strong and your heart rate not too slow. I do not, as a rule, give verapamil together with beta-blockers or digitalis (used in the treatment of cardiac rhythm disturbances and heart failure) for fear of dropping the cardiac rate too much.

Finally, almost everyone with angina should take *an aspirin a day* to keep the blood flowing within the coronary arteries unless there is some reason not to do so (bleeding or active peptic ulcer, very high blood pressure, or sensitivity to the drug). I recommend the 80 mg "baby" dose.

Sudden worsening of angina that has been stable for months or years is an important danger signal that requires evaluation. Tell your doctor about it immediately. The hospital is the best place to assess such "unstable" angina. If after a day or two of rest and adjustment of drug therapy you continue to have chest pain, chances are you will be sent for an angiogram.

How dangerous is an angiogram? That depends on your age, how strong or fragile you are, and most important, on the skill and experience of the doctors doing it. On the average, 1 patient in 2,000 may suffer a stroke, a heart attack, or die as a result of the procedure. But that's a lot less risky than not having this test when you need it. Before you sign the angiogram consent form, you'll be subjected to the legally mandatory "truth in advertising" requirement. You will be informed of *every single possible* risk and disaster that can befall you. After the intern or resident has recited to you all the dire consequences, he or she will blandly ask for your permission to proceed! More than one of my own patients have left the hospital against medical advice after being read this riot act! But before packing your bag and going home, remember that the actual risks are, in fact, minimal, as long as the procedure is done by a well-trained team with lots of experience.

There are three coronary arteries—one on the right (without a name of its own!), one on the left (the circumflex), and one going straight down the middle of the heart (the left anterior descending). The right coronary artery is a single vessel, but the circumflex and left anterior descending split off from a common trunk called the *left main coronary artery.* When that trunk is more than 50 percent blocked, I recommend immediate surgery (regardless of the severity of symptoms) because were it to close, a substantial portion of the heart muscle would suddenly be deprived of blood, resulting in a massive heart attack or death.

If the angiogram shows "high grade" disease in two or more arteries, and something needs to be done (because you're having angina despite all the medication you're taking), the first choice should, if possible, be angioplasty (ballooning). But don't kid yourself. This procedure is not really much less dangerous than a bypass, although it is obviously easier to do, involves virtually no pain, requires only a day or two in the hospital, and is considerably less expensive.

Neither ballooning nor surgery is a cure. They simply buy time for the collaterals to develop. After a successful procedure, you must not let up trying to control the risk factors that caused the arteriosclerotic blockage in the first place. Unless you do, you're a candidate for closure of the arteries that were ballooned open, or obstruction of the new bypass grafts, or progression of disease in those blood vessels that were normal at the time of your operation.

The risk of dying from bypass surgery is less than 2 percent nationwide, but higher if you are very sick, or if your heart muscle was severely damaged by previous heart attacks. How long will the new grafts remain open? About 20 percent close within the first three to six months (the closure rate after ballooning is about 30 percent in that same period of time). Does bypass surgery prolong your life, or merely control your symptoms? I think it does both.

In summary, angina is a chronic condition. Although you can lead a virtually normal life for many years on medication alone, you may need active intervention like balloon angioplasty or surgery. But this more aggressive approach should not usually be undertaken unless medical therapy has failed, or at least two of your coronary vessels are severely diseased.

Anxiety and Depression

Too many of us resort to and then become dependent on "mood-altering" drugs when confronted by life's problems. None of these agents provides a permanent solution, or even a long-lasting one. Instead of spending a little extra time to listen and to counsel someone who's upset, doctors are altogether too quick to prescribe a tranquilizer. I also deplore the indiscriminate administration of sedatives to the elderly, especially those in nursing homes, in order to make them more "manageable." But there are times when psychotropic drugs are necessary and helpful. In an acute crisis or depression, the appropriate medication can tide you over, and asking for it is not a sign of moral weakness. Here are some of the agents available for this purpose, how they should be taken, and under what circumstances.

Anxiety

The most important rule in prescribing (or taking) "something" for "nerves" is to appreciate the difference between *sadness* and *anxiety*. Agents that calm you down can worsen depression and those that stimulate you can increase an inner sense of stress. Unfortunately, the distinction between these two polarized emotions is not always made, even by doctors. Here, for example, is an anxiety scenario. Someone you love may be on a plane that has crashed. There *are* survivors, but the list is not yet available. As you sit waiting at the airport, you become progressively more anxious. Some of us can cope in such a situation, but others require—should be given and not hesitate to take—an *anti-anxiety* agent. Nor need the crisis be a serious one to generate a pervasive sense of doom. Some patients become nervous wrecks just waiting for an operation or other procedure. They too may require, and benefit from, pharmacological as well as moral support. Now, no medication is going to change the news you finally get about the airplane crash, or make your surgery unnecessary, but an anxiolytic (anti-anxiety drug) will control the temporary *symptoms* of anxiety that you may experience—the shortness of breath, palpitations, dizziness, and sweating—and allow you to function more effectively for "the duration." I wouldn't deprive any of my patients of such medication any more than I would withhold pain medication after surgery.

The *benzodiazepines* are the most effective anti-anxiety agents we have. The best known in this group are diazepam (Valium) and chlordiazepoxide (Librium). However, in older persons I prefer alprazolam (Xanax) because the body metabolizes (gets rid of) it more quickly, and chronic use does not result in its accumulation in the liver.

Remember, don't take *any* benzodiazepine for longer than 4 weeks if you can possibly help it, and when you decide to stop, don't do so abruptly or you will experience troublesome withdrawal symptoms—increased anxiety, insomnia, and so forth. Never forget, too, that most anti-anxiety agents are habit-forming.

Buspirone (BuSpar) is a newer anxiolytic with which I have also had some very good results. It comes in 5 mg tablets, and I recommend starting with 3 a day. If you need more, you can raise the dose by 5 mg a day every 2 to 3 days, but you should never take more than 12 tablets a day (60 mg). This drug may require up to 2 to 3 weeks to work, so if you need an immediate effect, as is often the case with anxiety, BuSpar won't do you much good. A big plus for it, though, is that it does not impair sexual function or slow you down. Since it is not a benzodiazepine, it doesn't result in withdrawal symptoms once it's stopped, or interact with alcohol like Valium does.

There is a special kind of anxiety—the *panic attack*—that has nothing to do with bad tidings. These crises arise suddenly, out of the blue. They leave you feeling threatened, perspiring, dizzy, and nervous; the heart pounds, and you experience a sense of suffocation. Before assuming that these symptoms are purely psychological, check them out with your doctor to make sure they are not the result of some physical disorder like hyperthyroidism (see page 264) or mitral valve prolapse (see page 190). Panic attacks that recur regularly can usually be prevented or at least reduced in frequency by specific therapy. I prefer fluoxetine (Prozac), 20 to 40 mg a day, and/or Xanax, 0.25 mg 2 or 3 times a day, for this purpose. Unlike many other psychoactive drugs, Prozac does not usually result either in weight gain or in a sharp blood pressure drop when you stand up quickly from the lying or sitting position. In fact, many of my patients report some loss of weight with it (but don't take it for that purpose)! Another plus for Prozac is that it does not produce significant drowsiness. Although there have been isolated reports of suicide and other adverse effects in persons taking this drug, in my experience they occur less frequently than is the case with other medication. However, like all drugs, Prozac and Xanax can produce side effects, so make sure that your doctor monitors you closely while you take them.

Stage fright is a specific form of panic attack that plagues even seasoned performers, politicians, and public speakers. This symptom responds best to propranolol (Inderal), 10 or 20 mg about 30 minutes before the "challenge." But try it first on a day you're not performing just to make sure you can tolerate it. There's no point in being so totally relaxed that you are unable to perform!

If you are afraid of heights or crowds (*agoraphobia*), have severe *claustrophobia* (for example, in an elevator), or experience anxiety symptoms when you find yourself in a specific situation over which you have no control, Valium, 5 or 10 mg, will tide you over. One of my patients who boarded a Concorde airplane for the first time had no idea the passenger cabin was as confining as it is. Her claustrophobia was so severe, she refused under any circumstances to continue the flight and forced the return of the aircraft to the terminal just as it was about to take off.

Whatever the reason, if you find that your anxiety crises are coming on with greater frequency and for fewer obvious reasons, you need more than medication. You should arrange for psychological evaluation, and if necessary, counseling.

Depression

Depression is quite different from anxiety. If someone you loved was killed in that plane crash, news of which you were anxiously awaiting, you're now apt to be depressed. Depression takes all the joy out of life; friends who formerly brought you pleasure now leave you completely "cold"; things you loved to do are no longer fun; there is nothing to which you really look forward; life is gray. Sometimes there's a very good reason for the depression:

You may have lost a loved one, been fired from your job, or had some serious financial reversal. As is the case with anxiety, no medication will convince you that any of these misfortunes was in reality a bit of good luck, but the right one can help you to carry on—and to sleep. (Most insomnia is due to depression and anxiety.) But before taking any antidepressant, make sure that your depression is not the result of some underlying illness or medication. For example, hypothyroidism slows you down—you look and feel depressed; beta-blockers (for hypertension or angina), the benzodiazepines (used in the treatment of anxiety), and even antibiotics can all deprive you of the *joie de vivre*.

The "best" antidepressants over the years have been the tricyclic drugs, the most familiar of which are imipramine (Tofranil), amitriptyline (Elavil), amoxapine (Asendin), and trazodone (Desyrel). Prozac is also useful in depression, as it is in panic attacks. The usual dosage is 20 to 40 mg a day. You may have to try several of the agents before you find the best one, but don't be impatient: It may take weeks before these drugs yield their best results.

Some persons become depressed, or manic, or both, in a cyclical fashion. They suffer from what is called a manic-depressive disorder for which the medication of choice is Lithium. The optimal dosage depends on your response and the concentration of the drug in the blood (which can be measured in a routine blood test that should be done at regular intervals). One of Lithium's complications is reduced function of the thyroid gland, which should be checked every 6 months.

In summary, if you need help because you're anxious or depressed, ask for it. That assistance should come in the form of *psychological support* from friends, family, your doctor, or a trained specialist, supplemented when necessary by drug therapy. Don't take any such medication for longer than is necessary to tide you over the critical period in your life. Some persons do require long-term therapy for their emotional disorder—schizophrenia, phobias, panic attacks, or manic-depressive disorders. *Every medication should first be tried in its lowest dosage and continued for the shortest period possible*. Remember that regardless of their benefits, these agents are for the most part potentially habit-forming and all of them have some side effects. And remember too that many drugs interact with each other, so be sure to tell each of your doctors what the other is prescribing.

ASTHMA, BRONCHITIS, AND EMPHYSEMA: You'll Huff and You'll Puff, But You'll Never Blow the House Down!

Every breath you take makes its way down the bronchial tubes, which, like the branches of a tree, become smaller and smaller until they end in the lungs as tiny air sacs. This is where the oxygen is extracted from the inhaled air and delivered to myriad small blood vessels within the lung tissue. These merge with the rest of the bloodstream, in which the oxygen is carried to every organ in the body. After the oxygen has served its purpose and been "used up," carbon dioxide, its most important by-product, is transported back via the veins to the lungs where it is exhaled and eliminated.

The act of breathing, of which we are normally unaware, is hard work for someone with lung disease. Depending on the type of pulmonary problem present, such individuals may have trouble getting oxygen down the bronchial tree and into their lungs, or there may be something interfering with its transfer into the circulation, or the carbon dioxide may not be freely eliminated and may accumulate in the body. Here is how best to treat each of these possibilities.

Apart from acute infections such as pneumonias of various kinds (see page 218), the three main disorders that affect respiration in an ongoing way are asthma, chronic bronchitis, and emphysema. All three cause obstruction to the flow of air in and out of the lungs and interfere with the exchange of oxygen between the lungs and the blood. During an *asthmatic attack,* the larger air passages (bronchi), which were perfectly normal only a moment ago, *suddenly* become narrowed by spasm, leaving you acutely short of breath and wheezing. But you bounce back to your normal state as soon as the acute attack is over. By contrast, in *chronic bronchitis,* the airways are continuously inflamed or diseased, and are filled with mucus, pus, and other junk. Affected individuals are constantly coughing, spitting, and short of breath, symptoms that become worse during flare-ups. Unlike asthma or bronchitis, the main problem in *emphysema* is not the air passages but destruction of many of the little air sacs from which the inhaled oxygen enters the bloodstream. Those that are left increase in size in order to compensate for the reduction in their number. So in emphysema, there is a chronic body-wide deficiency of oxygen because there are fewer air sacs through which it can be transferred to the blood, and the lungs are difficult to move because of the large air "blisters." That's why people with emphysema are short of breath.

In real life, asthma, bronchial disease, and emphysema are not separate, distinct conditions. They can and often do blend into one another. In general, asthma is not due to cigarette smoking, and does not lead to emphysema. However, chronic bronchitis and emphysema are often the result of heavy cigarette smoking.

Treatment goals of any lung condition are (a) to prevent the bronchial tubes from going into spasm, (b) to relax the spasm quickly when it does occur, (c) to control (prevent or eradicate) bronchial infection, and (d) to increase the oxygen supply.

Let's first consider the management of spasm in *acute asthmatic attacks*. At least 20 million Americans, mostly children and younger adults, suffer from asthma. In recent years, however, this disorder has been more frequently observed in elderly persons, too. The wheeze of the typical attack is easy to recognize, and is due to sudden spasm of basically healthy *large* bronchial tubes. This constriction is usually caused either by *allergy,* in children, or, in older adults, by a *hyperreactive* response of the respiratory passages to some trigger or other. Allergic attacks are usually set off by an environmental inhalant like pollen, dander, or house dust while hypersensitivity of the airways more often occurs after infections like bronchitis or sinusitis. At any age, an attack can be set off by a variety of factors such as sudden exposure to cold, irritant fumes, a burst of exercise (especially in children), and certain foods (particularly shellfish), preservatives, sulphites, and medications such as aspirin and beta-blockers.

It's important to end the asthmatic attack as quickly as possible regardless of what caused it. The best way to do that is with a *beta 2 agonist*. This medication relaxes and dilates the bronchi, allowing air to move freely in and out of the lungs. There are several beta 2 agonists on the market, and they're all pretty much the same. In order not to clutter my mind with lots of names to remember, I usually prescribe albuterol (Ventolin, Proventil). Beta 2 agonists come in three forms—aerosol, tablet, and injectable. An aerosol dispenser provides an exact and constant metered dose with each push of the button, acts within 5 or 10 minutes, reaches its peak effectiveness in an hour, and continues to work for 5 to 7 hours. Take 2 puffs as soon as the wheezing starts, and 2 more every 20 to 30 minutes for the next 2 hours if the spasm persists. If you're still wheezing after that time, you may not be using the aerosol dispenser properly, so when you are given a prescription for *any* aerosol medication, make sure you understand the operating instructions. Ask your pharmacist to show you the right way to use it, and to give you a dummy unit (no reflection on your intelligence) on which you can practice. This is how it should be done. First, take a deep breath in and blow it all the way out. Then, place the end of the nozzle in your mouth. Now, as you begin to breathe in slowly, depress the button on the unit. This releases the medication in a mist, which you inhale as you complete the breath. After you do, close your lips and don't breathe for about 10 seconds. This allows the drug to penetrate deep inside your air passages.

It's really quite easy, but some patients, especially children and the elderly, just can't seem to get the hang of it. If you have trouble coordinating your breathing, and are not getting all of the mist into your respiratory tree, there is a *nebulizer* available that will solve the problem. It's simply a longer dispenser attached to the nozzle of the standard unit. With the extended portion *in your mouth,* depress the release button in the usual way and inhale as much of the medication as you can. Because this unit has a one-way valve, any of the active medication left behind in the extended nozzle can be inhaled during your next breath. Although these nebulizers are bulkier, not quite as convenient to carry around, and somewhat more expensive, they do overcome the mechanical problems some persons have with the conventional dispenser.

In order to reduce the frequency of their attacks, I advise my patients with *recurrent* asthma to inhale Ventolin 3 times a day, followed each time by an aerosol *steroid.* There are several of the latter on the market, but I usually prescribe beclomethasone (Vanceril). As with other aerosols, remember to hold your breath for 10 seconds after each puff. Although they are generally well tolerated even when taken for months at a time, aerosol steroids are not always without side effects. Some of my patients have developed fungus or yeast infections in the mouth or throat, or become hoarse after prolonged use. To reduce that likelihood, you should rinse your mouth very thoroughly after each inhalation. Unlike the beta agonists, aerosol steroids do not provide immediate relief of the acute attack itself, but decrease the overall frequency of attacks by reducing inflammation in the airways.

If Ventolin and Vanceril do not prevent the asthmatic attacks, you'll have to move on to oral *theophylline.* There are several different preparations of this drug on the market, but I have found the Uniphyl brand especially convenient because it is slowly absorbed and its effect lasts for 12 hours. Take

1 pill in the morning and 1 at night. Be very cautious about using *any* theophylline product if you have liver disease or heart trouble, and never do so without consulting your doctor. You should also be aware that certain other drugs, like erythromycin or the anti-ulcer agent cimetidine (Tagamet), can interact adversely with theophylline.

Another agent, ipratropium (Atrovent) is a good backup for asthmatics. It won't terminate an attack like Ventolin does, but will reduce the severity. Atrovent comes only in aerosol form, and the dose is 2 to 4 puffs, 4 times a day immediately after using the beta 2 agonist. I recommend it for most older patients, providing they do not have glaucoma. It is otherwise well tolerated and virtually free of the intestinal and cardiac side effects sometimes associated with other anti-asthmatic medications.

Occasionally asthma is so severe, frequent, and resistant to treatment as to require *oral* steroids. These hormones are so dramatically effective that I have had patients call me, angry that I hadn't prescribed them sooner. The problem with steroids by mouth or injection is the price you pay for the relief you get if you use them for long periods of time: osteoporosis, high blood pressure, facial hair, peptic ulcers, cataracts, fluid retention, susceptibility to other infections—and these are only some of the possible complications. So, by all means, take steroids if necessary *to weather an acute asthmatic crisis*, but try to avoid them on an ongoing basis. In acute cases, I prescribe 60 mg of prednisone tablets (Deltasone) the first day, and then reduce the dose every couple of days so that by day 8 you're all done.

Asthmatic attacks, especially those brought on by exercise, exposure to animals, or other specific allergic situations, can sometimes be prevented, especially in children, by cromolyn (Intal). However, this agent is of no use in the acute attack. An oral antihistamine-like medicine, ketotifen (Zaditen) is now available in Europe for the prevention of asthma and initial reports about it are glowing. It has not yet been approved in the United States.

Here's a tip if you're asthmatic *and* absent-minded, and have left home without your medication. Caffeine is a mild bronchial dilator, so if you feel an attack coming on, two cups of coffee (the real McCoy, not the decaffeinated stuff) may tide you over until you can get some Ventolin.

A word about asthma and pregnancy. One third of all asthmatic women who become pregnant remain free from attacks during the nine months of gestation; in another third, the frequency is unchanged; in the remaining third, the asthma develops for the first time during the pregnancy. Every asthmatic woman should maintain close contact with and be guided by her doctor with regard to the medication she takes during her pregnancy.

When the Lung Disease Is Chronic

Most asthmatics are completely healthy *between* attacks, and usually enjoy normal exercise tolerance. One of my sons who is asthmatic was a member of the Yale swimming team! But if you have a cough that has lasted for at least 3 months a year for the last 2 consecutive years or more, are constantly hacking and spitting up, or are short of breath after minimal exercise or at rest, you probably have bronchitis, one of the *chronic obstructive pulmonary diseases* (COPD). The odds are you're also a cigarette smoker, 15 percent of whom end up with COPD.

Unlike uncomplicated asthma, chronic bronchitis is a disease of older persons. The earliest symptoms are chronic cough and shortness of breath, which gradually worsen with time. You're apt to blame it all on your cigarettes if you're a smoker, or your extra weight if you're fat, or on being "out of shape" if you're lazy. If you lose weight, work out, or quit smoking, you will feel better—for a while. But most patients become progressively more short-winded and continue to spit up globs of grayish mucus, especially in the morning. In many, the course is slowly downhill. They are able to tolerate less and less exertion until finally they are short of breath even at rest because the diseased air passages are now chronically inflamed or obstructed.

When chronic bronchitis is complicated by emphysema, large amounts of air are trapped in the lungs, expanding them and leaving the chest barrel-shaped. Such "lungers" generally purse their lips when they exhale in order to slow breathing and allow more air to be exhaled with less effort. In far advanced COPD, stagnant air accumulates in the lungs and results in less available oxygen, as well as in the retention of waste gas (carbon dioxide).

Pulmonary disease that has reached this stage is incurable, but the right treatment can slow down its progress and make you feel better. However, no therapy is worth a hoot unless you *stop smoking*. Each and every puff you take from your cigarette, pipe, or cigar further reduces the amount of oxygen available to you.

At this advanced stage you need daily medication to widen the constricted air passages as much as possible. I usually prescribe an aerosol beta agonist agent like Ventolin, 2 inhalations every 4 to 6 hours, followed by Atrovent, 2 puffs, 15 or 20 minutes later. I supplement this basic treatment with oral theophylline and aerosol steroids. If you have a heart condition, your doctor should keep an eye on your heart rate and blood pressure, both of which can be increased by excessive use of the beta 2 agonists.

If you have COPD, every "cold" is a major threat to you. Unless you abort it, the cough becomes bad, and the mucus you spit up turns yellow or green. So at the very first suggestion of a respiratory infection in such individuals, I prescribe trimethoprim-sulfamethoxazole (Bactrim DS) twice a day for 10 days (except for those who are sulfa-sensitive) and every winter I give most of my COPD patients tetracycline, 250 mg, 4 times a day on Saturdays and Sundays. Also, because prevention and treatment of infection are so important, everyone with COPD should have an annual flu booster and a once-in-a-lifetime pneumonia shot.

At some point, as the COPD becomes more severe, you are likely to need oxygen—first when you overexert yourself, then during the night, and finally on and off throughout the day. How much to take, and for how long, is determined by analyzing your blood for its oxygen content. Dependence on oxygen doesn't mean you're at the end of the road. I have several patients and friends who for years have required it for as long as 18 hours a day and remain productive, though physically inactive. They go out socially and even shop, but always carry a portable oxygen supply.

Here's the scoop on some really "hot" new research in emphysema. A small percentage of patients with this disease suffer from a genetic disorder in which there is a deficiency of an enzyme (alpha 1 antitrypsin) needed to maintain the integrity of the little air sacs at the end of the bronchial tree. You may be one of them if you have emphysema and are still in your forties or younger, aren't a smoker, and especially if other members of your family also have this disease. The missing enzyme has now been synthesized and is already available. If you turn out to be deficient in it (something that can be determined by a blood test), replacement may slow the progression of your disease.

BALDNESS: A Remedy for the Lucky Few

Baldness in both sexes is sometimes due to disease. For example, you may lose your hair if you have dropped a great deal of weight in a short period of time, have had general anesthesia, are receiving certain anticancer drugs, are recovering from an infection with high fever, were taking birth control pills for years and then stopped because you wanted a baby, have "tired blood" due to an iron deficiency anemia, or if your thyroid function is either too low or too high. In these circumstances, the hair comes out in bunches; there is more than the usual amount on your pillow in the morning or after you brush and comb it. However, it does grow back when the underlying cause is eliminated.

Most cases of baldness, however, are *not* due to disease but are genetic and, unfortunately, permanent. Such inherited or "pattern baldness" affects many more men than women. Twenty-five percent of the male population begin to be bald by the time they reach age thirty, and about two thirds are either bald or have a balding pattern (in which the hair recedes from the forehead and top of the head) by age sixty. So we're talking about 20 to 30 million individuals in the United States alone. This inherited tendency appears to be transmitted more often from the *maternal* side of the family, which is what I emphasize to my three sons, all of whom are losing their hair. It's interesting how differently these "children" of mine deal with their "condition." The eldest capitalizes on it. In fact, he actually shaves what little hair is left on his head. When I ask him, "Why are you doing this, Yul?" he tells me that sparse head hair is evidence of his virility, and he further reminds me that "eunuchs never go bald." My middle son is too involved with molecular biology to pay much attention to his lack of hair. To him it's just another personal characteristic like the color of his eyes. However, he does compensate for what he lacks on his head with a beautiful, luxuriant beard (the genes for hair loss do not affect the face). My youngest son, until very recently, spent most of his extra money on every kind of advertised and nonadvertised potion touted to restore hair growth. He feels about baldness much like Caesar did when he asked the Roman senate for permission to wear a laurel wreath around his balding head.

Unlike "toxic" baldness, the onset of pattern baldness is gradual and subtle. The average scalp contains about 100,000 hairs (blondes have as many as 140,000, brunettes 155,000, and redheads only 85,000), of which we normally lose 100 per day. Each individual hair survives for an average of about 4½ years, during which time it grows about half an inch a month. In its fifth year, it falls out and is replaced within 6 months by a new one. But if you're destined to become bald, when that "old" hair dies, it is not replaced. In other words, genetic baldness is not due to excessive hair loss, but rather failure to produce new hairs.

Until recently, the only really effective way to prevent baldness in the genetically vulnerable was castration. But times have changed. Scientists at the Upjohn Company some years ago noted that patients taking minoxidil, a preparation used to lower elevated blood pressure, often grew hair where least expected or desired. This was particularly vexing to women (although hair on the forehead is not particularly attractive in men either). The commercial potential of this unwanted side effect was not lost on these investigators, who set about developing a topical preparation of minoxidil, which they hoped might be useful for the treatment of pattern baldness. They found that a 2 percent solution worked, and have marketed it under the name Rogaine. (The FDA vetoed "Regaine.") Best results are obtained in young men in their early twenties and thirties who are just becoming bald, although there are more and more success stories being reported in older persons—and women, too. Rogaine makes a dramatic difference in only less than 10 percent of cases, but it does reduce the progression of baldness somewhat in most men and women with the genetic form. Apply it to the scalp twice a day for at least one year before deciding whether or not it's going to help. It takes that long to be sure. (Incidentally, that "experiment" will cost you about $600, and the drug requires a doctor's prescription.) If you are not impressed after 12 months, stop the Rogaine. Perseverance beyond that time is futile and expensive. If, however, you do see a difference, you will need to continue it for the rest of your life. Adding topical retinoic acid (Retin-A) in the 0.025 percent strength to the scalp once a day along with the Rogaine may yield even better results. So Will Rogers was not entirely correct when he said, "The only thing that can stop falling hair is the floor."

Most of those for whom I have prescribed Rogaine have tolerated it well. One man, however, grew hair on his ears, while a few individuals complained of headaches, dizziness, weakness, and even impotence. One report from Canada tells of patients who experienced an increase in heart rate and other cardiac symptoms, none of which appeared to be harmful. I understand that higher concentrations of minoxidil are now being evaluated, with results that appear to be more impressive than those observed with the 2 percent strength. But these studies have not yet been completed.

I am very grateful to the Food and Drug Administration for stating officially that Rogaine is the *only* medical treatment that can influence hair growth, and that all the other lotions and potions, pills, and herbs are worthless. That will save my youngest son and many of my patients a good deal of money!

If Rogaine doesn't affect your baldness, and your appearance really bothers you, *punch graft hair transfer* will leave you less bald. Tufts of hair taken from the back of your head (which is immune to genetic influences) are transferred up front where it counts. Patients of mine who have been "punched" in this way tell me that it hurts, but is worth it!

There is an even more extensive and effective surgical approach called *scalp reduction* in which the back rim of the head together with its hair is pulled up to the crown. One of my patients just paid $15,000 to have this procedure done—without insurance reimbursement. It works, and it is usually permanent.

If you're not happy with the "bald look" and can't afford the more expensive procedures, there's always the option of "cosmetic correction." A good wig will fool almost everybody.

Flash! Here's some late-breaking news as we go to press. I sent my youngest son a 6-month supply of Rogaine as a birthday present. Two months later he has developed a peachlike fuzz where he used to be totally bald! He's absolutely thrilled. I was going to delay publication of this book until there was more to see, but my editor advised against it. Hopefully, I'll have some "before" and "after" pictures for the *second* edition.

BITES: When Dog or Man Bites Man

Every year, two million people in the United States go to their doctors or hospital emergency rooms because they were bitten. In 85 percent of cases the attackers were dogs—their own or someone else's (most commonly a pit bull or a German shepherd), in 10 percent they were cats, and the remaining aggressors were rodents, other animals, and ladies and gentlemen!

Most human bites are the result of interpersonal violence. (Who knows how many occur among "consenting adults" in circumstances too embarrassing to describe?) You may have read about people with AIDS who have deliberately bitten others in order to infect them with the disease. One such incident involved a prisoner in an American jail who was sentenced to twenty additional years for biting a guard with the intent to kill. I've never heard of anyone actually contracting AIDS by this route, but if these biters continue trying, it's going to happen one day.

The major complication of a Homo sapien's bite is infection. It's likely to be more serious than an animal bite because the specific mix of bacteria in the human mouth, especially in the dental plaque, tends to be resistant to antibiotics. So if you're in a fight with someone, and land a punch in the mouth, and they happen to have teeth, you may be in more trouble than your victim. Infection from a bite can remain limited to the skin's surface, extend to deeper tissues. If a bone or joint is affected, the result can be arthritis or osteomyelitis (an infection of bone). Puncture wounds are the worst.

If your skin has been broken as a result of a bite, what best to do depends on its depth, location, and what you know about the biter—animal or human. But in every instance, first clean the area as best you can, using whatever antiseptic you have at home. Then quickly go to an emergency room for more sophisticated attention. You will, of course, need an antibiotic, and the best one in such cases is amoxicillin (Augmentin), 500 mg, 3 times a day for as many days as the wound is open or draining. Although you may require stitches, it is generally better for these bite wounds to heal without being sewn up. This prevents any dangerous organism from being locked into your body. The exception is the face, where unless the wound is closed "cosmetically," you may end up permanently disfigured.

It's important to know something about the person or animal that bit you because in addition to local infection, you can contract hepatitis B and herpes from humans, and rabies from an unknown dog, raccoon, or other wild animal. If there is any question about the source of the non-human bite, be sure to get the rabies vaccine; 25,000 Americans do every year. To prevent herpes, I suggest a 5-day course of acyclovir (Zovirax) in a dose of 200 mg, 5 times a day (see page 129). If there is a chance of having been bitten by someone with hepatitis, get a hepatitis B immune globulin shot immediately, and have yourself vaccinated, too (see page 127). If you're due for a tetanus booster anyway, this is a good time to have it.

If the "biter" is known to have AIDS, you've got a real problem on your hands. Of course, you'll clean the wound immediately and thoroughly, and apply whatever antibiotic ointment is handy. But that's not going to kill the AIDS virus. In all honesty, I can't think of anything specific to do, except keep checking your HIV status every few weeks. Chances are the bite didn't "take."

BOWEL CANCER: Surgery—The Only Cure

If you are American, there's a one in twenty chance that someday you will develop cancer of the colon. That likelihood is even greater if this malignancy runs in your family. No cancer is "good," but unlike that of the pancreas, lung, or kidney, cancer of the bowel can often be cured, and there are usually enough clues to make early detection possible. *However, such cure can only be obtained by the complete surgical removal of the cancer.* Chemotherapy and radiation may help buy time and control symptoms in advanced cases, but will not permanently rid the bowel of the malignancy.

Because finding bowel cancer early on can so often result in its eradication, it makes good sense for you to follow the guidelines offered by the American Cancer Society.

• Everyone over fifty years of age should have their stool tested *chemically* for the presence of blood (not always visible) at least once a year.

• When you reach fifty, your bowel should be examined by means of a sigmoidoscope, a thin flexible tube with a light and forceps at its end. This makes it possible for the doctor to look into the colon and remove for biopsy any suspicious tissue that is present. If two such examinations one year apart are normal, then you needn't have another for 5 years (unless you or two or more close blood relatives have had bowel cancer or polyps; in that case, have the flexible sigmoidoscopy done every two years). Some "cost-conscious" analysts would have you believe that this test does not save enough lives to justify the expense. Most doctors who deal with people, not numbers, disagree.

In addition to the American Cancer Society guidelines, I recommend that every routine physical after the age of forty include finger examination of the rectum. This may detect a cancer low down, in both sexes, and prostatic malignancy in men.

If, in the course of a bowel examination, you are found to have a "growth," don't panic. Chances are it is a benign polyp. Once it's all out, you're cured. Sometimes, however, what *seemed* like a benign polyp to the naked eye turns out to be malignant when examined under the microscope.

If you have cancer of the bowel, the *entire* growth must be surgically removed, if possible. Once that has been done, the question often is raised whether any other treatment is required. For example, should you also have *radiation* just in case some cancer cells were left behind? Is there some *chemotherapy* you should be taking for good measure as well? Except for rectal cancer, radiation will make no impact on your survival, but chemotherapy can. I am convinced, from all I've seen and read, that it's a good idea to have additional anticancer treatment for 6 to 12 months after the operation. The combination of 5-FU (5-fluorouracil) and levamisole is remarkably free of side effects, and seems to improve survival even in advanced cases.

Bowel cancers that are found too late to cure spread first to the liver, then to the lung. We used to think that once they reached the liver, the outlook was hopeless. Not so. If a scan indicates only a single lump or nodule (metastasis), cure is still possible if that focus can be completely removed by an operation. Two of my own patients remain free of disease four and five years after such surgery on the liver. But metastases to the lung are uniformly fatal.

I'm particularly interested in the possibility of using a vaccine in the treatment of bowel cancer. After the malignant tissue has been excised, it is "deactivated" by irradiation, and injected back into the patient. These "dead" cancer cells may be able to stimulate the body's immune system to produce antibodies that, at least theoretically, can destroy or neutralize any live tumor cells that were left behind. This research is currently being pursued, and the data are still incomplete, but it may hold promise and is worth knowing about.

In summary, remember, the key to the cure of colon cancer is early detection, and *immediate* removal of the polyp or tumor. Follow-up chemotherapy and immunotherapy, even if the malignancy *appears* to have been totally cut out, have also been shown to be effective. The field of cancer treatment is ever-changing. You can learn about the latest developments by calling the Cancer Hotline at 1-800-4-CANCER. It's free and available to everyone.

BRAIN SEIZURES: They're a Symptom, Not a Disease

On a recent flight from New York to Dallas, while I was having a casual conversation with a young woman seated next to me, she suddenly, and without warning, started to flail about, jerking her arms and legs, foaming at the mouth, finally lapsing into semicoma. The diagnosis was no mystery: this was a typical "seizure." All I could do at 28,000 feet and 500 miles an hour was to lay her down (not so easy given the "generous" seating that airlines now provide) and make sure she didn't hurt herself. The attack ended after a few minutes, and aside from looking somewhat dazed, the young lady seemed none the worse for it; neither did she have any recollection of what had happened. It was as if she had simply awakened from sleep.

Seizures such as this vary in severity, ranging from what appears to be nothing more than a little "daydreaming" to a violent fit. They are usually due to a sudden burst of abnormal electrical activity from some irritable focus in the brain, caused by any one of a number of different conditions—scar tissue from an old injury, a stroke, heart disease, kidney trouble, sinus infection, middle ear disorders, high fever (especially in children), chemical abnormalities in the blood, and acute alcoholic toxicity or other drug poisoning. When a clear-cut mechanism responsible for recurrent seizures is not apparent, even after extensive study, we assume it's either some birth injury to the brain or a metabolic disorder.

Although most people use the term *epilepsy* in describing these attacks, a more modern designation is "seizure disorder." However, when the "fits" occur regularly, we still call them *epileptic*.

If you ever witness a seizure yourself, try to record *exactly* what you saw (how it started, what limbs were affected, how long it lasted), and tell the patient's doctor all about it. An accurate description of what transpired can be of great help in determining the proper diagnosis and treatment. But in addition to a good history, complete physical exam, and blood tests, special investigation is also usually required—an electroencephalogram (think of that as an electrocardiogram of the brain), CAT scan, and magnetic resonance imaging (MRI).

The most important approach to seizures is to prevent them, and that can be done in 90 percent of cases, especially if a specific cause has been identified. The antiseizure drug I prefer for adults is enteric-coated valproic acid (Depakene). The required dose varies from person to person, and an adult may need from 1,000 to 3,000 mg a day. Valproic acid can cause liver problems, so be sure to have blood tests at regular intervals while you're on this medication. When valproic acid pushed to its dosage limit fails to prevent seizures, I advise my patients to switch to or add one of the following agents: phenytoin (Dilantin), carbamazipine (Tegretol), or phenobarbital. Dilantin can give you a rash, swollen gums, and neurological problems. Although it's probably the best agent for children under age two, it can, nevertheless, lead to mental deficiency in infants; Tegretol may be carcinogenic after prolonged use, can seriously affect the blood, and may also result in double vision and impaired balance; maintenance therapy with phenobarbital can induce hemorrhages in infants. Whichever medication works best for you with the fewest side effects should be continued for at least 5 years after your last seizure.

BREAST LUMPS: When They're Cancerous

There is nothing more frightening to a woman than a lump in her breast, which in one of every eleven American females turns out to be malignant. You may have found it yourself as the vast majority do; your doctor may have picked it up; or it may have been diagnosed on a routine mammogram.

In order to detect breast cancer early enough to cure it, the American Cancer Society recommends that all women have a screening mammogram starting at age forty, then every 2 years for the next 10 years and annually thereafter. I personally also advise that the first one be done at age thirty-five, especially when there is a family history of breast cancer. In addition, you should also examine your own breasts every month, and ask your doctor to do so, too, at each routine visit to his or her office. Routine mammography is particularly important for women over age sixty-five who have never been pregnant, because their risk for developing breast cancer is two to three times that of all other females.

Once a lump has been found, regardless of your age, family history, what you or your doctor *think* it is likely to be, and even what the mammogram suggests, you should insist on a "tissue diagnosis." Deciding whether or not a growth is malignant simply on the basis of its size, appearance, location in the breast, or feel, is not enough. There isn't a doctor I know who hasn't at some time or another been fooled by such "impressions." There's only one *sure* way to tell, and that's by examining under a microscope the tissue obtained from the lump—in other words, having a biopsy done. Depending on where in the breast the lump is situated, as well as its size, that biopsy can be performed either by needling the growth and sucking up some of its contents, a procedure called *fine needle aspiration* done in the doctor's office, or by incision, performed in a hospital operating room. The needle technique is less desirable than the surgical biopsy because of the danger of "false-negative" results (the tiny needle simply doesn't "find" the cancer).

The best news you can get after a biopsy is that the lump is "benign." But what should you do if it is malignant? Until only a few years ago, it was standard practice to remove the entire breast, the glands in the axilla (armpit), and all of the muscles and tissues of the chest wall as well. This was called a *radical mastectomy,* or the Halsted operation (after the doctor who first proposed it). But then specialists took a second look at the survival statistics over the years in women who had had various kinds of surgery for breast cancer. They found that the mutilating operation was no more effective in the long run than less drastic approaches. And so the radical mastectomy was replaced by the *modified radical mastectomy,* in which the breast and the glands in the armpit are removed, but the remaining tissues in the chest wall are spared. Although less disfiguring, this operation is still emotionally devastating to many patients. More and more women are now asking whether *lumpectomy,* in which only the tumor itself is excised and the rest of the breast is spared, is adequate. Statistics from thousands of cases indicate that a thorough lumpectomy *followed by X-ray treatment* to the remaining breast tissue results in exactly the same survival rate as the more extensive modified radical procedure—certainly after five years, and probably at ten. Reliable data are not yet available for any longer than that.

In order for lumpectomy to be feasible, the tumor should be no larger than 2 inches across, should not involve the nipple, and the breast must be free of any other suspicious areas. Lumpectomy leaves the breast with a much more satisfactory cosmetic appearance so that additional reconstruction is not usually needed. I recommend lumpectomy and radiation to all my patients whose cancer satisfies the criteria listed above. The follow-up X-ray therapy consists of a 2-minute treatment every day for 5 weeks.

Regardless of which procedure is done, lumpectomy or modified radical, the surgeon will remove and analyze the glands located in your armpit (axilla) in order to determine whether the tumor has spread to involve them. This information has a vital bearing on your subsequent treatment. Even if the cancer is present in these lymph glands, all is not necessarily lost (although it is obviously better for them to be free of disease).

After its removal, the cancerous tissue is examined to see whether or not its cells "bind" or take up the female hormones, estrogen and progesterone. On the basis of this analysis, the tumor is designated as "positive" or "negative" binding. This determines what additional treatment may be required to reduce the risk of recurrence and prolong your life. Since 1957, scientists in several countries have been conducting cooperative studies to evaluate the benefits of chemotherapy and hormonal therapy, and which patients should receive what treatment. On the basis of their pooled information to date, there is now general agreement that:

- If your cancer is *hormone-receptor positive,* regardless of whether or not it has spread to the lymph glands, your survival is improved by taking an anti-estrogen drug called tamoxifen (Nolvadex) for at least 2 years after the surgery. In fact, I personally recommend that it be continued indefinitely despite some of its side effects—weight gain, clotting disorders of the blood, and gynecological tumors, so make sure you're examined appropriately at regular intervals if you stay on this medication.

If the hormone-positive cancer has spread to your lymph nodes, then, in addition to the tamoxifen, you should also receive chemotherapy for one year. The most effective regimen consists of four drugs—cyclophosphamide (Cytoxan), methotrexate (Rheumatrex), doxorubicin (Adriamycin), and 5-fluorouracil. Never refuse chemotherapy for fear of its side effects; there are ways to control them. A year seems a long time to be taking such drugs, but the prolonged survival they confer is well worth it.

- If the tumor is *receptor-negative,* you do not need tamoxifen, but should have chemotherapy—for 6 months if the glands of the armpit were free of cancer and for at least one year if they were not.

In summary, always consult a breast cancer specialist before deciding on how your malignant lump should be treated. Keep "abreast" of the latest developments in this field. The statistics and recommendations are constantly changing as more data from ongoing studies become available. At the moment, there are definite "protocols" that spell out who should receive either hormones or chemotherapy, or both, and for how long, after the breast cancer has been surgically removed. Every woman with such a malignancy should know what these recommendations are and discuss them with her doctor. Again, remember the Cancer Hotline—1-800-4-CANCER.

BREAST PAIN: The Kind That Comes Back Every Month

Many premenopausal women suffer from "*cyclical mastalgia*"—that is, pain, heaviness, and tenderness in the breasts that recurs every month like clockwork just before their periods. (Don't confuse this disorder with cystic mastitis, in which the breasts are lumpy and uncomfortable throughout the *entire* menstrual cycle.)

Until recently, all you could do for cyclical mastalgia was to take conventional pain killers. However, bromocriptine (Parlodel), a hormone that also helps some patients with Parkinson's disease, effectively prevents this particular kind of breast pain. So if you dread your next period because of the breast discomfort that heralds it so predictably, ask your doctor about a maintenance dose of bromocriptine—2½ mg, once or twice a day. If you continue this therapy all month, there is a 75 percent chance that you will not experience cyclical mastalgia with your periods. If you have an aversion to pills and decide to stop the bromocriptine, you may get a bonus pain-free interval for the next 3 to 6 months. You can then decide whether to resume the drug or simply to take pain killers as needed.

The side effects of bromocriptine, when they occur, are nausea, vomiting, and dizziness, all of which disappear when the drug is stopped. After menopause, when the production of hormone (estrogen) responsible for cyclical mastalgia is reduced, the condition clears up and you no longer need either bromocriptine or analgesics.

BREATH ODOR: When Your Friends Will Neither Tell You Nor Kiss You

Some healthy people have persistent, offensive breath regardless of what they eat, the toothpaste they use, or the mouthwash they swish. I remember one childhood friend whose presence I could always detect at a party just by the "aroma" in the room. It wasn't a matter of oral hygiene or garlic and onions, but simply his unique "breath signature."

If you've had "normal" breath all your life, and suddenly develop halitosis, it's usually due to some specific condition—diseased gums, an abscessed tooth, a throat infection, sinusitis, a lung infection, Sjögren's syndrome (in which the eyes and mouth are chronically very dry), chronic alcoholism, or a variety of disorders such as kidney failure (toxic substances that should have been excreted by the kidneys are retained and impart a urinary smell to the breath), or liver disease (in which there is a "mousy" mouth odor). Many individuals have offensive breath because of what they eat—raw onions, garlic, cabbage, and the like. Vitamin supplements are another source of bad breath. If you've ever opened a bottle of B complex vitamins, you know what I mean—the "aroma" is enough to make you gag. A steady diet of these vitamins, especially in megadoses, can cause chronic halitosis, and impart a characteristic color and odor to your urine as well.

Some duodenal ulcers are infected by a bacterium called *Helicobacter pylori,* a bug that not only keeps the ulcer active, but may also cause halitosis. In fact, some specialists believe that it can inhabit the intestine without causing ulcers—just foul breath. Theoretically, antibiotics could cure this form of halitosis by eliminating the infection. But the diagnosis is not easy to prove. A biopsy of the duodenum is required to identify the organism positively. I can understand submitting to that if you've got an ulcer that won't heal or keeps recurring—but for halitosis?

So what should you do for your bad breath? Obviously, if it's the result of one of the underlying causes above, you'll have to eliminate it. Masking the odor with a mouthwash is no solution. In fact, the combination of halitosis and some of these "antiseptics" can yield a more offensive aroma than the bad breath itself. What's more, frequent use of medicinal mouthwashes with a high alcoholic content may be associated with a higher incidence of oral cancer, especially in persons who also smoke and drink to excess. The ads would have you believe that chewing chlorophyll lozenges will purify your breath regardless of what you've eaten, drunk, or smoked, but I've never found them particularly effective. If you need to take something, fresh parsley or a strong mint are probably the most effective measures available.

BURNS: A Matter of Degree

No one knows for sure how many people are burned every year in the United States, but about 2 million such victims visit emergency rooms or doctors' offices for treatment. Seventy thousand require admission to the hospital, many are left permanently scarred, and 10,000 die. The sad thing about these terrible statistics is that 90 percent of all burns are the result of carelessness—and avoidable.

There are several different kinds of burns: *electrical* burns, which injure structures deep in the body but barely affect the skin surface; *chemical* burns, from acid or some other substance spilled on the skin; and *heat* burns.

On the news the other day I heard about a fire in which a local dignitary had been badly burned. The announcer reported that 28 percent of the man's body surface was involved. I suddenly realized that I wasn't quite sure what that meant. How does one quantify body area? What is a 28 percent burn? I looked for the answers in an old surgical textbook that I hadn't used since I was a medical student, and came across a diagram of the body that resembled the charts hanging in a butcher's shop. It was divided up into little squares containing a lot of different percentages; there was no way I would ever remember them. But later I came across the "rule of nines," which makes it all very simple. Each upper limb constitutes 9 percent of the body surface; each lower extremity 18 percent; the head and neck total 9 percent; the front of the body equals 18 percent, and the back of the torso is 18 percent. That adds up to about 99 percent, which is close enough. If you want to estimate the total amount of a burn that is patchy in distribution, an area covered by your palm and fingers involves 1 percent.

In addition to the *size* of the burn, its *depth* is also important. Is it first, second, or third degree? A *first degree* burn, the kind you might get when you've had too much sun or accidentally touched something very hot, affects only the most superficial layer of the skin, the epidermis. It's painful and red, but *does not blister.* The redness is due to dilated blood vessels on the outermost layers of the skin, which blanch when you press on them. After a couple of days, first degree burns usually begin to peel. *Second degree* burns are deeper and extend below the epidermis. They *usually blister* immediately or within 24 hours. The burned skin is raw, pink, wet, and very sensitive, so that any medication or ointment you apply is going to hurt. As a matter of fact, even just blowing on the area may be uncomfortable. A second degree burn takes about 2 or 3 weeks to heal if it has not been infected, *and does not usually leave a scar.* However, the original skin pigmentation may never fully return. *Third degree* burns affect the deepest layers of skin and cause extensive blistering. Unlike the more superficial injuries, the burned skin is neither pink nor red but almost white, as if it had been cooked, which is, in fact, precisely what has happened! The nerves in the involved area have been damaged so that it's actually less sensitive, it takes up to 2 months or longer to heal, and almost always leaves a scar.

Now that you know how to classify a burn, and the different types, here's how they should be treated. If you have a first degree burn, run cold tap water over the affected area for about 20 or 30 minutes, and then apply cold compresses. That's much better and not nearly as messy as the old-fashioned honey, butter, and other "recipes." To reduce the sensitivity of the injured skin, apply a topical ointment or cream between applications of the cold compresses, and cover the area with a light dressing. My favorite such product is Cortaid with aloe vera, which you can buy without a prescription. Aloe vera has been used for centuries as a medicinal agent. It's not only very soothing, but it may prevent infection and reduce the amount of peeling. You may, if necessary, also apply Nupercainal ointment or cream for additional local pain relief (to which some people are, or become, allergic). If aspirin doesn't control the pain, a half to one grain of codeine prescribed by your doctor almost certainly will.

In contrast to first degree burns, you can't just put any ointment or cream on a *second degree* burn, because the exposed area under the blisters *absorbs* many topical substances. So, after the cold water rinses, it's best to see your doctor, who is likely to apply silver sulfadiazine on the skin (unless you're sensitive to sulfa) and cover it with a light dressing.

The treatment of *third degree* burns has greatly improved in the last few years; it includes sophisticated antibiotic therapy, replacement of lost body fluids, and grafting of artificial skin when necessary, all of which are best obtained in a burn unit that can provide the necessary facilities and specially trained staff. This is exemplified by the Shriners' hospitals throughout the country, which provide such care to burn victims under the age of eighteen free of charge.

CANKER SORES: Easing the Pain

Cankers are not a threat to life, but they're painful and a nuisance. Unlike herpes ("fever blisters," or "cold sores"), cankers are not due to a virus but are probably the result of a temporary malfunction of the immune system. You can tell a canker from a herpes lesion by its appearance and location in the mouth. Herpes frequently involves the skin of the lips, cankers do not; herpes usually strikes the gums, cankers rarely do; herpes sores are often accompanied by tender glands in the neck, not so for cankers; cankers have a slightly raised yellow border surrounded by a reddish area, herpes sores are like ulcers.

Cankers are particularly common in children who wear dental braces, but can afflict anyone for no apparent reason (women more often than men). You'll know you have one even without seeing it when you wake up one morning and can't find a comfortable place for your tongue, it hurts to eat and to chew, and most fruits and juices, like the orange juice you have for breakfast, sting like the devil.

What is the best way to treat a canker? In my experience, it will take its own sweet time to disappear no matter what you do! When I was a boy, I used to call my dentist whenever I had one. Regardless of what he prescribed, it took 2 weeks to clear up. He knew as much (or little) about curing cankers then as I know now! The best advice is to avoid salty, spicy, and hard foods that can irritate the sore. But I have also stumbled on a "remedy" you can buy without a prescription. Although not a cure, it eases the pain and seems to heal the cankers a little more quickly. It is called Zilactin—a mixture of tannic acid and alcohol in a gel that forms a protective covering on the canker when applied directly 2 or 3 times a day. If Zilactin isn't enough, ask your doctor to prescribe an oral rinse of 2 percent lidocaine viscous, which you can swirl around your mouth every few hours. In severe cases, rinsing with topical steroids after meals may be necessary.

CERVICAL CANCER: Is It, or Isn't It?

There's probably no routine screening test for cancer that generates more confusion and anxiety than the Pap smear of the cervix. For in addition to the 20,000 definite cases of cervical cancer discovered every year in this country, there are many, many more in which the diagnosis is "maybe later, but not yet"—reflecting the presence of "atypical" cells under the microscope. (Doctors call it "dysplasia.") I don't think anyone knows for sure how many of these "precancerous" lesions actually progress to the real thing if left alone. Estimates vary between 5 and 70 percent! (Dr. Myron Buchman, who reviewed this section, opts for 40 percent.) But it's important for you to know that dysplasia is common, that there are several grades of severity, and that it requires treatment.

The cervix, the lowest portion of the uterus, juts into the vagina where it can be felt, seen, and examined by the gynecologist. Its lips surround an aperture that opens into a canal leading into the interior of the uterus. That's the route the sperm follows for its rendezvous with the egg. During pregnancy, the cervical lips become progressively softer so that they can dilate sufficiently when it is time for the baby to exit the womb. The cervix sees lots of action in the lifetime of most women, from the trauma of innumerable penile thrusts to a variety of infections. No small wonder it is so often the seat of cancer.

Cancer of the cervix is not the same entity as cancer of the main portion of the uterus. The locations, cell types, diagnosis, and treatment differ between them. Cancer of the cervix is most often diagnosed by a Pap smear, which involves removing a sampling of the cervical cells with a small brush and examining them under a microscope. A "negative" or "normal" result is no guarantee that you don't have cervical cancer. The malignant cells may have been missed by the brush, and yes, let's face it, the technician or doctor reading the slide may also have missed the diagnosis. So, accuracy of the test depends both on sampling and proper interpretation. Despite these variables, the Pap test is an indispensable routine screening technique in women without symptoms. But if you have *any* abnormal vaginal bleeding or spotting, whether it's after intercourse, or postmenopausal, or related to your periods, you should have *both* a Pap test *and* a curettage or scraping of the cervical canal.

When describing treatment options for any disease, I like first to present the salient facts about its *prevention.* You *can* reduce the risk of developing cervical cancer by controlling or avoiding the following factors that seem to predispose to it. For example:

• Shun *multiple sex partners.* You never know what cancer-causing agents a lover may bring to your cervix. It is especially important to be prudent in your pre- or early teens because that's when the cervix is probably "set up" or primed for cancer developing later in life. This malignancy appears to be transmitted in much the same way as venereal disease. For example, the papilloma virus, spread by intercourse, certainly does increase the risk of cervical cancer; genital herpes may also do so (components of the herpes virus have been found in cervical tumors); even the ubiquitous chlamydia organism has been implicated. The case for a causal relationship between infection and cervical cancer is further strengthened by the fact that women who use barrier methods of contraception, in which the cervix is physically protected, have a lower incidence of this malignancy.

• More women who smoke cigarettes have cervical cancer than do nonsmokers, although nobody really understands why. Their cervical mucus, and that of passive smokers, too, contains a large amount of nicotine. So if you're worried about getting cancer of the cervix (not to mention cancer of the lung, bladder, mouth or pharynx, or a heart attack), stop smoking, right now, and never start if you haven't already done so.

• If your mother was taking diethylstilbestrol (DES) during her pregnancy, you are at increased risk for developing *vaginal* cancer. I suggest you have a colposcopic exam every six months (the gynecologist introduces what is basically a small microscope into the vagina, through which he or she can see that portion of the genital tract in great detail, and remove samples from suspicious-looking areas).

• Beta-carotene, thought to be protective against lung and other cancers, may also reduce the incidence of cancer of the cervix, as may vitamin C and folacin (one of the B complex group).

Early *detection* is the key to the successful cure of cervical cancer. I recommend *yearly* pelvic exams and Pap tests for *all* women starting in their teens, with only a few exceptions, as noted below. I do not agree with those who tell you that every two or three years is often enough. That's a sop to cost-effectiveness and not good medicine. Nor is an active sex life a prerequisite for Pap testing. A woman who no longer engages in sex may have picked up a carcinogenic virus years earlier, during which time the cancer has been cooking, to become apparent only later when sex is just a memory!

• If you've been treated for cervical condylomas, the warts due to viral papilloma, I suggest a Pap smear two or three times a year for 2 years, and then annually.

• If you've had a hysterectomy in which the uterus has been removed but the cervix has been left behind, you need a Pap smear only every 3 years, not annually. Of course, if you've had a complete hysterectomy in which the cervix was also removed, then there's nothing to "Pap"!

• If your gynecologist tells you that your Pap smear looks "suspicious," ask what was found and what is planned for dealing with it. If dysplasia is present, you must consider it precancerous. When a patient asks me what to do about this finding, I suggest colposcopy. This permits the gynecologist to identify the abnormal site and biopsy it. A curettage or scraping of the cervical canal should also be done.

The best way to guarantee the eradication of abnormal cervical cells is by *conization,* the removal of the portion of the cervix in which they are located. But since this procedure may result in bleeding, infection, and perhaps reduce your chances of having a baby at some later date because of scarring, it should be reserved for severe cases. When dysplasia is mild or moderate, other methods such as freezing, laser vaporization, and burning (with a hot cautery) are easier and preferable. However, their long-term cure rate is a little lower. Whichever approach you choose, make sure to have repeat Pap smears every three months for at least 2 years just to make sure you're not one of the 5 or 10 percent of women in whom cancer subsequently develops.

If the Pap test reveals not dysplasia, but a very early malignancy that is still entirely within the cervix and has not affected the rest of the pelvic area, the best treatment here is surgical removal of the cervix (conization)—no ifs, buts, or whys. It's now too late for freezing, burning, and lasers.

If you failed to have routine Paps on schedule, or ignored abnormal vaginal bleeding, the doctor may find a cervical cancer that has spread to local glands or now involves the vagina, the wall of the pelvis, the urinary bladder, the rectum, and even remote sites such as the lung. In these advanced cases the options are *surgery* (if the cancer can still be completely removed), *radiation* (when the malignancy is too far gone to be excised), *chemotherapy* (after it has spread to distant parts of the body), or a combination of the three.

In summary, the incidence of cancer of the cervix can be reduced by a prudent sex life and the avoidance of sexually transmitted disease. It can almost always be detected by annual Pap tests in all females after puberty. There is an almost 100 percent chance of cure if it is discovered at an early stage. Left alone, cancer of the cervix can kill you.

CHICKEN POX AND SHINGLES: One and the Same Disease

The names we give some of the "ordinary" diseases of childhood often puzzle me. Take chicken pox, for example. It has absolutely nothing whatsoever to do with chickens—you don't get it from chickens, you don't give it to chickens, and you won't acquire it from eating chicken. The scientific name, varicella, is no more enlightening—all it suggests to me is a pasta dish.

Until very recently, chicken pox was the only childhood illness for which there was no vaccine. One has now been prepared and is expected to be available soon. The disease usually runs a benign course in children, but can cause problems in adults. It can be recognized by the characteristic *blisters* on the face, scalp, and trunk that come and go in crops over a 7-day period.

Since it's better to get over and done with chicken pox when you're young, we don't normally isolate children with this infection. In fact, we actually encourage them to mingle with most of their friends, since anyone who has already had the disease (and 90 percent of Americans have by the time they reach adult life) is immune to reinfection. But we do keep them away from *adults* who have never had it, from *pregnant women* whose past chicken pox history is uncertain and who may be vulnerable to infection, and from anyone whose *immune system* is impaired (individuals with AIDS, those who have received organ transplants and are taking immuno-suppressant agents to prevent rejection, and patients receiving cancer chemotherapy).

What therapy should a child with chicken pox receive? There's no drug that destroys the virus, so all you can do is make the patient comfortable. Since the major symptom in chicken pox is itching, everything possible should be done to control scratching because that leads to chicken pox's main complication—infection of the skin blisters. The most effective way to minimize the itch is with the antihistamine trimeprazine (Temaril). The dose depends on the child's age and weight, so check it out with the pediatrician. Cool compresses and calamine lotion (straight, without phenol), available without a prescription, are also very soothing to the skin. Should any of the blisters become infected, an oral antibiotic such as erythromycin or cloxacillin should be used in doses based on the child's weight. To lower fever and reduce the malaise, give acetaminophen (Tylenol), *not aspirin.* (*Remember!* Aspirin is associated with a higher incidence of Reye's syndrome, a severe neurological disorder, when taken by youngsters with flu, chicken pox, and other viral diseases.)

If inflammation of the brain (encephalitis) sets in, as it may on rare occasions, Dilantin (phenytoin sodium) will control the accompanying convulsions. Remember, however, that encephalitis, though frightening, clears up in a day or two in the great majority of cases.

In adults, the big worry with chicken pox is *pneumonia,* because it is resistant to all antibiotics. The best and only treatment is acyclovir (Zovirax), the antiviral drug that is so effective in treating herpes. For chicken pox pneumonia, which is potentially life-threatening, the acyclovir is given either intravenously or in large oral doses, 20 of the 200 mg capsules a day for at least 10 days. It represents an important advance in the treatment of this condition.

If you are pregnant and have been exposed to chicken pox, the chances of your giving birth to an abnormal baby are less than 5 percent—not great, but real enough. If the diagnosis is confirmed, you should receive an injection of *chicken pox immune globulin,* available from The American Red Cross. This won't necessarily reduce the chances of your baby being born with some congenital abnormality, but it will render your own infection less serious. The greatest danger is in contracting chicken pox late in the pregnancy—that is, near the time of delivery. Should that happen, the infant, not the mother, should immediately be given the chicken pox immune globulin injection, for it may otherwise die.

Shingles

Shingles is caused by the same varicella virus that results in chicken pox. After the initial chicken pox infection, the virus migrates from nerve cells on the skin's surface, along their fibers, and ends up in the nerve relay stations on the outside of the spinal column. There it lies dormant—and just waits. Years later, the virus becomes reactivated, after some emotional stress, or an infection, or following the use of medications that compromise the immune system, or often for no apparent reason. This "born-again" virus then retraces its steps along the nerves, reemerging on the surface of the skin where its travels began.

Shingles almost always affects only one side of the body—that is, its symptoms and rash *rarely cross the midline.* An attack begins as burning, itching, or pain in a discrete area of the body, followed after a few days by the appearance of a rash that closely resembles that of the original chicken pox.

Shingles is treated in two ways. The first objective is to relieve the pain. Second and perhaps more important is the prevention of *postherpetic neuralgia*—severe pain in the affected area that persists for months and sometimes years after the telltale rash has completely cleared. For control of the pain of the acute attack, I recommend calamine lotion with 1 percent phenol applied locally. Codeine, in a strength of a quarter or half grain by mouth, every 4 hours, will relieve severe pain. *But the most important treatment of all is oral acyclovir* (*Zovirax*) in high doses—800 mg, 5 or 6 times a day for 10 days. Not only does this drug reduce the severity of the acute shingles attack, it also decreases the chances of persistent pain after the rash has cleared. Should shingles involve your eyes or ears, I recommend *steroids* along with the Zovirax as soon as the diagnosis is made, in order to reduce the chances of permanent damage. This is administered by a physician either orally or by injection.

If you are unlucky enough to be left with residual pain after your rash has cleared, there is now a topical substance that you can apply to the affected area. It is marketed as Zostrix, and is made from an extract of red pepper called capsaicin. It has afforded some relief to several of my own patients.

CHLAMYDIA INFECTIONS: The Most Common Sexually Transmitted Disease of All

Almost five million Americans harbor chlamydia, a more recently recognized player among the infectious agents that cause sexually transmitted disease in the Western world. AIDS, syphilis, gonorrhea, and herpes may be household words, but chlamydia infects more men and women in this country than all of these combined. In fact, if you have had syphilis, gonorrhea, or herpes, chances are you're also playing host to chlamydia, for which you should be tested and treated (except if you're a male homosexual, since chlamydia is not transmitted from man to man). If you've been infected with HIV, the presence of chlamydia can accelerate the appearance of AIDS symptoms.

Chlamydia infection in women causes vaginal irritation and a yellowish-green discharge; in men, it results in a whitish-mucoid urethral secretion, pain in the scrotum, and discomfort on urination.

Special tests are necessary to identify the chlamydia organism because it's so tiny and hard to see under the microscope with the usual staining techniques. But it should always be looked for because unless eradicated by antibiotics, it leaves you chronically infected—pelvic inflammatory disease in women, epididymitis in men (infection of the tubules that carry the sperm from the testes where they are made, to the urethra whence they are ejaculated)—and infertile. Should a female harboring chlamydia become pregnant, her infant is likely to develop either pneumonia or a serious eye infection in the first 6 months of life, as well as asthma and permanent lung damage later on.

The best treatment for genito-urinary chlamydia is doxycycline (Vibramycin, Doryx), 100 mg, twice a day for 7 days, or tetracycline, 500 mg, 4 times a day. Check the bottle to make sure the drug has not expired. Don't settle for some capsules that were prescribed years earlier and have been lying around the house. Buy a fresh supply, as much as it hurts to throw the old stuff away, because expired tetracycline and doxycycline can hurt the kidneys. Avoid these antibiotics if you are pregnant; they can cause permanent yellow, gray, or brown discoloration of your infant's teeth and retard growth later in life. Use erythromycin instead, but remember that the various formulations are not equally well tolerated. I prefer the EES brand (erythromycin ethylsuccinate) in a dosage of 4 of the 400 mg coated tablets a day for a week. A coating on this preparation protects the antibiotic against the action of food or milk in your stomach, so you may take it at evenly spaced intervals independent of what and when you eat.

CHOLERA: Is the Vaccine Worth Taking—And What's the Best Treatment If You Don't?

Cholera is not a major problem in the "developed" nations. This discussion will, however, be of interest to anyone planning to travel to Asia, Africa, or South America, where the disease remains a significant health hazard.

The organism responsible for cholera (*Vibrio cholerae*) is transmitted by food and water contaminated by excrement. It causes a toxic protein to attach itself to the cells that line the intestinal tract, resulting in a kind of diarrhea you wouldn't believe! The fluid literally pours out! This tremendous loss of water and minerals leads to shock and collapse of the cardiovascular system. Therein lies the threat to life. So the *immediate treatment* of cholera is the intravenous replacement of solutions that contain the various substances excreted from the bowel in these huge amounts. Such a preparation recommended by the World Health Organization consists basically of salt, potassium, sugar, and bicarbonate, and is available wherever cholera is endemic. But if you're stuck out in the boondocks where no one has this particular WHO formula, you can make your own mixture while waiting for more sophisticated intravenous therapy. Add 2 tablespoons of molasses and a teaspoonful of salt to every liter of water (boiled, naturally), and drink as much of it as you can—until it virtually comes out of your ears. *Do not take any of the antidiarrheal medicines* like Lomotil, Imodium, codeine, or Kaopectate; they do no good in this particular situation and may actually hurt you.

Together with the fluid replacement, start the *specific treatment* for cholera—tetracycline, 500 mg by mouth, 4 times a day. It works like magic! In only 3 days, the diarrhea will stop. Like the American Express card, never leave home without this antibiotic if you're going to any part of the world in which cholera is present. Since tetracycline stains the teeth of young or yet-to-be-born children, pregnant women or persons under eight years of age should take furazolidone (Furoxone) instead, 100 mg every 6 hours for 3 days.

The treatment of cholera is so simple, and the cure so predictable, that I do not usually recommend the vaccine to my own patients. In my opinion, the discomfort and cost are not worth it (although the shots themselves are not dangerous). However, if you're going to an area where an epidemic is actually in progress, and the authorities insist that you be vaccinated, you'd better comply. If it is optional, pass it up.

CHOLESTEROL: You *Do* Have a Fat Chance!

Hundreds of millions of dollars have been spent on cholesterol research over the years, and many scientists have devoted their entire professional careers to studying the relationship between blood fats and vascular disease. I was a member of the original National Heart Institute Task Force on Arteriosclerosis in the early 1970s, and together with my twelve colleagues on that panel, spent almost 3 years evaluating the "lipid hypothesis"—the proposition that arteriosclerosis and cholesterol were causally related. Based on all the available evidence *at that time,* we did not feel we could justify across-the-board restriction of dietary cholesterol as a matter of national policy. We did, however, strongly recommend that additional large-scale studies be conducted to help resolve the issue. Today, some 20 years later, much of this research has been completed. On the basis of these new data, the "medical establishment" in the United States, led by the government's National Heart, Lung and Blood Institute, the American Medical Association (AMA), and the American Heart Association (AHA), has officially taken the position that too much cholesterol in the blood does, in fact, contribute to hardening of the arteries, and that lowering it not only helps prevent arteriosclerosis but can even result in its regression. Although there are still some dissenting voices in the scientific community, they are very much in the minority.

I believe that it is now reasonable to conclude that cholesterol is *one of several* factors responsible for hardening of the arteries in *some people,* perhaps even in most. *It is, however, not the only cause.* And what troubles me is that the cholesterol levels we are all being advised to achieve are unrealistic and almost impossible for most people to maintain for any length of time. If we focus on a single number rather than on evaluating the entire biochemical profile of each individual, many of us may end up unnecessarily using anticholesterol *drugs,* whose long-term effects are not as yet known.

So here's what I tell my patients when they ask what I think about the "cholesterol controversy":

- Do not be intimidated by either those who totally deny the importance of cholesterol (yes, there are still some doctors who do) or the enthusiasts who overemphasize it. The truth is that while too much does no one any good, how much is "too much," and for whom, is still up in the air.
- Don't deprive your kids of *all* fat and cholesterol while they are still growing and developing their vital organs. Overzealously withholding all dietary fats and cholesterols may hurt them at this stage in their lives.
- Life is too short for you to eat like a rabbit. A "prudent" diet like the one described below is enough to keep you healthy.
- Don't take any cholesterol-lowering medication unless the blood level is above 250, *and* is accompanied by an LDL (the "bad" cholesterol) greater than 160, and/or an HDL (the "good" fraction) below 40. But if your blood picture is "bad" and you suffer from angina, have had a heart attack, or there is a family history of premature heart disease, then you should certainly be on some anticholesterol agent.
- Don't force a restrictive diet on someone of seventy years or older. There are more important nutritional problems than cholesterol to worry about at that age. Many senior citizens can't get enough to eat for economic reasons, or their dentition is so bad that they can't chew what is available, or they're depressed, can't see, or can't cook for themselves for other reasons—Parkinson's disease or a stroke—the list is very long.

If your cholesterol level is "high," and you've been advised to lower it, make sure the laboratory numbers are accurate before starting any kind of dietary intervention or medication. There should be at least three readings (preferably from different laboratories), always including not only the cholesterol level itself but the HDL and LDL figures too. A cholesterol value may be high but "permissible" if the LDL (the bad fraction) is low, and the HDL (the protective component) is way up. If the cholesterol/HDL ratio is less than 4.5, you're probably in the clear. You can raise the HDL and further reduce that ratio by eliminating cigarettes, exercising regularly, and having one or two drinks a day (except if you have a drinking problem).

If repeat testing confirms the need for some action on your part, begin with your diet. It's not easy, it requires a lifelong commitment, and despite the most rigid adherence, it doesn't always work. For example, a strict vegetarian may have a high cholesterol level because of a metabolism that makes large amounts from the "building blocks" in the blood. On the other hand, a fat and cholesterol glutton may have "ideal" levels because his or her body is "burning up" the excessive amounts being consumed. But you'll never know into which category you fall until you at least *try* to follow a diet. Here are some general guidelines.

Focus on decreasing your consumption not only of cholesterol, but also of fats, the worst being the saturated variety present in meat and dairy products. If you won't give up red meat entirely, then at least make sure all its visible fat has been cut away, and that your steak is not marbleized (the most delicious, delectable, tender, and deadly kind there is). Remember, too, that animal organs like liver, kidney, and sweetbreads are especially rich in cholesterol.

Fish is low in saturated fat, so have at least two servings a week, especially the deep-sea, cold-water variety like bluefish, tuna, salmon, mackerel, and sardines. They all contain the omega-3 fatty acids that protect against heart disease. Shellfish used to be forbidden, but they no longer are. They happen to be extremely low in saturated fat, and some, like mussels, scallops, and oysters, don't contain much cholesterol either. The consensus these days is that you may have all the oysters you want, but limit your intake of crustaceans (shrimp, lobster, crawfish, and crab) to three servings a week because of their higher cholesterol content. Of course, never have them with drawn butter!

Poultry is okay provided it's skinned. Eggs are the richest source of cholesterol. A single yolk contains 300 mg (the maximum allowed for the entire day in the diet recommended by the American Heart Association). So limit your intake to four a week. Eat all the egg white you like because it has no cholesterol and is fat-free, but avoid butter, cream, and hard cheese (if you're a cheese lover, try ricotta, mozzarella, or low-fat cheese). Skimmed milk is the only kind to drink. And while on the subject of "drinking," if you love coffee, prepare yours any way you like except by boiling. It seems that boiled coffee significantly raises cholesterol—especially the harmful LDL fraction.

How you prepare your food is almost as important as what you eat. Steaming, baking, or rack roasting, in which the fat drips away, are the preferred techniques; deep frying is out. Don't be fooled by the labels on cooking oils. Just because a product is "vegetable" doesn't mean it's good for you. Though triumphantly labeled as "cholesterol free," it may contain lots of saturated fat, which is even worse! As a matter of fact, at my house we do not use cooking fats that are solid at room temperature because they're all saturated. You're much better off with monounsaturated olive oil or canola oil made from rapeseed, both of which actually lower cholesterol somewhat.

Lots of fruit, vegetables, and *soluble fiber* are good for you. Not everyone understands what fiber is all about. There are two kinds: the *insoluble* form that doesn't dissolve in water, and the *soluble* one that does. Insoluble fiber (like all-bran) acts on the *lower bowel;* it makes your movements bulky and soft, it prevents constipation, and it may reduce the risk of colon cancer, but it doesn't do much for cholesterol. Since Mr. Kowalski wrote his *Eight Week Cholesterol Cure* book, oat bran has become the best-known *soluble* fiber. However, it's no different from any of the others in this category, like beans, dates, apples, figs, prunes, broccoli, or cauliflower, all of which lower cholesterol to the same extent as does oat bran.

Many people I know take lecithin supplements, a complex fat whose richest natural source is egg yolk, in order to improve their cholesterol numbers, but I have not been able to find any convincing evidence in the scientific literature that it does so.

If your cholesterol level remains unacceptably high after 6 months of conscientious dieting, you're going to have to start thinking about medication. But before beginning *any* drug, try psyllium, a concentrated form of soluble fiber found in such bulk laxatives as Metamucil and Konsul. It is safer than any medication, its side effects are minimal, it costs a lot less than prescription drugs, and 1 teaspoon, 3 times a day, in conjunction with a diet, can drop the cholesterol by as much as 35 percent, depending on how high the level was to begin with.

A new coated garlic pill from Germany, with a name that sounds like the call letters of a radio station (KWAI), has been reported to lower cholesterol levels by as much as 15 percent. Try it; you may like it. The manufacturer says your breath and skin will not smell bad.

If diet, psyllium, and garlic do not produce the desired result, and you are advised to take drugs, remember that it's usually a lifetime commitment. The agents don't cure the problem, they simply control it. The moment you stop taking them, the cholesterol levels bounce right back. Although most of the anticholesterol agents currently available *appear* to be safe, the long-term effects of some are as yet unknown. What's more, taking any such drug does not mean you should abandon your diet. It is not a license to eat all the wrong foods—eggs, thick, juicy marbleized steaks, whole milk, cream, and butter. While an effective pill will drop your cholesterol level, it's apt to do so much more if you continue to follow a prudent diet.

I usually start my patients who need medication on *niacin* (nicotinic acid). It does all the right things—it lowers cholesterol, raises HDL (by about 20 percent), and lowers LDL (by approximately 30 percent). Niacin comes in many different strengths and formats—elixirs, capsules, and timed-release spansules. I do not recommend the latter because even though they are better tolerated, they do not increase the HDL levels as much as the noncoated preparations do, and they are much more likely to hurt the liver. But whatever form you choose, *never break or chew niacin.* Swallow it intact because it will otherwise damage the lining of your mouth and throat.

In the important Coronary Drug Project Research Group Study completed some years ago, men who'd had a heart attack were treated for 9 years with a variety of drugs. Those who took 3,000 mg of niacin a day had significantly fewer recurrent nonfatal heart attacks, and a lower overall death rate. But this agent, especially in so high a dosage, has many side effects—flushing, itching, burning, tingling of the skin—because it dilates superficial blood vessels in the face, neck, and chest (so never take it together with hot liquids or alcohol, which have the same effect). These symptoms are more tolerable if you start with a small dose, like 100 mg, twice a day with meals. If the flushing is still a problem, then take an aspirin tablet about a half hour before each dose. Aside from leaving your face beet red (the most obvious side effect), niacin can also aggravate an ulcer and increase uric acid levels (so if you have a history of gout, keep away from niacin); it can raise the blood sugar level (for which reason I don't give it to my diabetic patients); and it not infrequently affects the liver, sometimes seriously (liver function usually returns to normal when the drug is discontinued). But niacin will often improve your blood-fat profile, and it is safe so long as you keep its potential problems in mind and are properly monitored to detect them while taking it.

Although the most effective dosage of niacin was shown to be 3,000 mg per day, I do not usually prescribe more than 1,500 mg. To give you some idea how "therapeutic" this amount is, the recommended dietary allowance (RDA) is only 18 mg for men and 13 mg for women! The price spectrum among the many generic and brand name preparations is staggering. For example, a month's supply of 1,500 mg per day of a generic form costs the druggist about $6 or $7; he pays more than $50 for the same amount in slow-release form and brand name.

If the largest amount of niacin you can tolerate doesn't correct your blood fats, add one of the *bile acid binding resins*. These drugs lower cholesterol by "binding" or combining with the bile acids in your gut, which are made in the liver and which are very rich in cholesterol. The resin hangs onto this cholesterol, and together they travel down the gut to be excreted in the stool. The liver then extracts more and more cholesterol from the blood in order to make up for what's being lost in the bowel. The net result is a drop in the blood cholesterol level—where it counts—and a reduction of the LDL as well (by about 20 percent). I like the fact that these bile acid binding drugs lower cholesterol without themselves being absorbed by the body. You swallow a dose, it enters the stomach, does its number on the bile acids, and leaves the body without ever getting into the bloodstream.

In a large study, one of these resins, cholestyramine (Questran, available either in powder form or as a "candy" bar), reduced the total serum cholesterol and the bad LDL, and slightly increased the good HDL. It also appeared to slow the progress of coronary artery disease in some patients, and when taken in a dose of 20 grams a day for 7 years, slightly lowered the death rate from coronary disease, too. Another resin with a similar action to Questran is colestipol (Colestid), which comes in granule form. I start my patients on 2 of the 4-gram packets of Questran a day, and increase the dose gradually to 6 daily (24 grams).

The bile acid binding resins should be taken *before* meals. Both the powder and the granules are not exactly delicious. Have you ever tried drinking a glass of sand? To make them more palatable, I suggest you mix them with about 3 ounces of a noncarbonated liquid. Try water first, and if that doesn't appeal to you, use unsweetened fruit juice or a thin soup. You can also mix them with cereal, applesauce, or crushed pineapple. Preparing the next day's supply the night before and refrigerating it makes it much more acceptable.

I've never had a single patient tell me they developed a taste for either of the bile acid resins. Instead, they regale me with tales of constipation, abdominal pain, and heartburn; they belch, they're often nauseated, and they are bloated. The constipation can be so bad, especially if you're predisposed to it in the first place, that unless you drink lots and lots of water, or use a stool softener, or eat plenty of fiber, your stool can become impacted and require extraction by the doctor. So let's all salute the 1,900 men who entered the study I mentioned above, who took 20 grams of this drug every day for 7 years!

If you're using a bile acid resin, also take a vitamin supplement that contains fat-soluble A, D, E, K, and folic acid, since these substances, like cholesterol, can also be bound by the resin and lost in the stool. It's because of this effect on nutrition that I do not recommend these drugs for children or pregnant females. Resins also bind other medications, so, for example, if you are on anticoagulants, be sure to check your prothrombin time more frequently. Also, take the Questran (or Colestid) either 4 hours before or 1 hour after any other drug just to make sure their absorption is complete.

If niacin and cholestyramine, singly and/or in combination, do not lower your cholesterol enough, then I'd switch to gemfibrozil (Lopid), 600 mg, twice a day, 30 minutes before breakfast and dinner. Like cholestyramine, it has also reduced the number of heart attacks in an impressive 5-year research project called the Helsinki Heart Study. Lopid not only lowers cholesterol, but raises HDL (by 9 percent) and lowers LDL (by 12 percent). I have always found it to be well tolerated. Side effects, when they occur, are mostly intestinal—cramps, diarrhea, or nausea—and they disappear when the dosage is reduced or the drug is discontinued. To maximize its effect, I usually combine it with niacin. Although Lopid was shown to be safe in the Helsinki Study, I still advise pregnant women and those who are breast feeding not to take it.

We now come to the most effective cholesterol-lowering drug of them all, lovastatin (Mevacor). Despite its effectiveness and relative freedom from toxicity, I prescribe it only after other regimens have failed. My reluctance to recommend it right off the bat reflects some lingering uncertainty about its long-term effects. Mevacor is available in 20 and 40 mg tablets. The starting dose is 20 mg with the evening meal, which can be increased gradually every 4 weeks to a maximum of 40 mg twice a day. Unlike the bile acid resins, this agent is absorbed by the body, and interferes with the manufacture of cholesterol by the liver. It also lowers LDL by as much as 30 percent. The two major short-term side effects to watch out for are liver damage, which clears up when the drug is stopped, accelerated cataract formation, and painful muscles. So, if you're on Mevacor, have your blood checked every 3 or 4 months not only to see how you are responding to treatment, but also to evaluate liver function. Have your eyes examined, too, twice a year. I've never seen an ocular complication in any of my own patients, but the reason for the precaution is that when the drug was in development, cataracts were observed in several Beagle dogs! Avoid Mevacor if you are pregnant, breast feeding, or under twelve years of age. Nor is it a good idea to combine it with Lopid, especially if you have any kidney or liver disease, because of the possibility of muscle damage.

A final word. In your enthusiasm to normalize your cholesterol and other blood fats, remember that this is only one step in the overall program to prevent arteriosclerosis, and that other risk factors, like high blood pressure, cigarette smoking, being overweight, and physical inactivity should be given as much attention as cholesterol abnormalities.

CIRCULATION DISORDERS IN THE LEGS: Living with a Charley Horse

Consider this: You're over sixty, probably a smoker, and may also have high blood pressure, diabetes, or heart disease. You've noticed that when you walk at a pace that used to make you feel good, you now develop a cramp in your calf (or in your thigh, or foot) after just a few blocks. You stop, do some window shopping, and the pain clears up in a couple of minutes. You're off again, but this time you can walk a little further than initially before the cramp returns.

Some patients ignore these symptoms and assume they are just due to "getting older." That's not so. The culprit is *arteriosclerosis* (hardening and narrowing of the arteries), not the aging process. Such vascular disease can also affect your brain (where, if severe enough, it can cause a stroke), your heart (leaving you with angina or a heart attack), your eyes (culminating in loss of vision), or your kidneys (which stop working properly so that waste products accumulate in the blood instead of being passed out in the urine).

When you describe the calf pain to your doctor, the first thing he or she will do is look at your feet. Are they blanched? Are there any sores that don't seem to be healing? If you're male, have you lost the hair growth on your toes? Then he'll feel the pulses. Are they reduced or absent? By this time, he'll have a pretty good idea whether or not your symptoms are due to a vascular problem. He may do some additional tests in the office to confirm his impression.

When given the diagnosis that the arteries in their legs are diseased and cannot deliver enough blood, most people worry that they may one day require amputation. That rarely happens, and when it does, it's usually in diabetics with arteriosclerosis of the arteries who have neglected to take proper care of their feet—who haven't been conscientious about keeping the skin clean, dry, and free of cuts and other injuries.

Here's what I recommend to anyone with vascular blockage in the legs:

- Stop smoking—immediately! That alone will help by eliminating nicotine-induced spasm in the already diseased arteries. No other treatment makes sense if you continue to use cigarettes.
- Walk for *at least an hour every day* on a level surface at a pace just short of that which causes pain. If you begin to feel discomfort, stop, wait until it passes, then walk some more. Unlike angina pectoris, caused by narrowed coronary arteries, leg pain from exertion is not dangerous.
- Take good care of your legs and feet. Any break in the skin should be cleansed immediately and a topical antibiotic applied. A limb that is poorly nourished due to a decrease in its blood supply cannot effectively resist infection.
- Avoid exposure to very hot and very cold temperatures.
- If you're diabetic, keep your blood sugar as close to normal as possible by following your diet carefully and taking the right amount of insulin or sugar-lowering drugs.
- If you're overweight, slim down. It's hard for your legs to propel a 250-pound mass of fat.
- If your cholesterol is high, lower it by diet and/or drugs. There is evidence that normalizing blood fats can cause the obstructing plaque to shrink and thus increase the blood flow to the area.

Will any medication help? Pentoxyphylline (Trental) is worth a try. It comes in 400 mg strength, and the dosage is 3 pills a day. Trental is the only agent approved by the FDA for the treatment of poor circulation in the legs. It renders the red blood cells more flexible so that when they reach a narrowed portion of the artery, they change their shape, squeeze through the obstruction, get beyond it, and deliver their load of oxygen on target. Don't give up on Trental until you've tried it for about 3 months. It may take that long completely to alter your red blood cells. Trental may not allow you to walk as far or as fast as you'd like, but it *will* increase, by about 40 to 50 percent, the distance you can go before pain sets in. There's no harm in trying it, especially since it has virtually no side effects.

Forget about drugs that are supposed to "dilate" your blood vessels. Sounds good in theory, but they don't work, and are a waste of money. They're ineffective because the diseased, rigid arteries cannot, in fact, dilate. These drugs only widen arteries that are already open, and so steal badly needed blood away from those that are blocked.

If possible, avoid or reduce the dosage of any beta-blocker you may be taking, because these medications can induce spasm of the already narrowed leg arteries. They also cause the heart to pump less vigorously, and so further decrease blood flow through the vessels. Another agent to keep away from is ergot, which prevents migraines by constricting the painfully dilated pulsating arteries in the head. Unfortunately, it also constricts blood vessels in the legs.

When should you consider a direct approach such as surgery, angioplasty, or laser therapy to relieve the blockage? Since arteriosclerosis of the leg arteries is rarely if ever a threat to your legs or your life, my advice is to continue a conservative program until your symptoms seriously disable you. When you can no longer walk any meaningful distance, or your legs begin to hurt at rest, then it's time to consider "invasive" forms of treatment to improve blood flow. But first, you'll need an angiogram of the lower extremities—in which dye is injected into the circulation in order to identify the exact location and severity of the blockages in your arteries. *Angioplasty* in the legs works exactly as it does in the heart. A guide wire with a collapsed balloon at its tip is introduced into the artery alongside the obstruction (it can't be done when the blockage is complete because then the balloon can't nestle beside the plaque). Once in place, the balloon is inflated, compressing the plaque against the wall—thus widening the vessel and increasing the blood flow through it. *Lasers,* which remain experimental for coronary artery blockage in the heart, are being used more and more to improve blood flow to the legs. They evaporate the obstructing plaque instead of pushing it aside. The laser is effective when the obstruction is total, whereas the balloon is not. Finally, there is *bypass surgery.* As in the coronary artery operation, the obstruction in the area is bypassed, usually by a graft made of Dacron.

If you decide on surgery, be sure to select an experienced team that performs this operation frequently and successfully. And before you sign on the dotted line, ask lots of questions, like the number of cases they do each year, and their results.

CIRRHOSIS OF THE LIVER: Not from Alcohol Alone

The liver is a low-profile organ with which most people are not emotionally involved. For example, you don't see posters in airports exhorting you to eat "liver-healthy foods" or which solicit financial support for a "crusade against liver disease." Despite the absence of hype, this organ is critically important because it performs a multitude of vital functions, and when diseased, can make you very sick or kill you.

Almost any injury, infection, or inflammation of the liver can lead to permanent damage and the replacement of normal healthy cells with nonfunctional scar tissue. If extensive enough, this process is called cirrhosis. Blood returning to the heart to be "recycled" is prevented by the scar tissue from passing freely through the cirrhotic liver and so backs up in the abdomen and the legs. This interference with normal flow is why someone with cirrhosis has a belly full of fluid (a condition called "ascites") and swollen legs. When the veins in the rectum are engorged, too, there is usually blood in the stool, and when those in the food pipe are distended, they may rupture and cause death from hemorrhage. In addition to blocking the return of blood to the heart, the liver can no longer efficiently perform many of its life-preserving functions. When that happens, the bleeding and clotting mechanisms it controls go awry; the normal breakdown of female hormones is interfered with so that estrogens accumulate in abnormal amounts (resulting in large breasts, lack of sex drive, and other female characteristics in previously sexy men) and jaundice often sets in. All in all, a very sorry picture.

In the developed world, cirrhosis is usually the end stage of chronic alcoholism. Elsewhere, it's most commonly due to previous hepatitis B or other viral infection. But we are now seeing more and more cirrhosis following such infection in this country, too. (If you've eaten raw clams and developed hepatitis A, don't worry—that virus rarely causes permanent liver damage.) Exposure to certain chemical toxins and drugs also accounts for a relatively small number of cases of cirrhosis.

Once the liver has been infected by a virus, there's nothing that will stop it from running its course. But toxins are a different story. *So, for example, if you are a problem drinker, and your blood tests reveal impaired liver function, you absolutely must stop drinking now.* Better late than never. Don't wait for New Year's Eve and yet another "resolution." The alternative to going on the wagon immediately is a slow, painful, and almost certain death. If some poison other than alcohol has given you cirrhosis, identify and eliminate it from your environment. Such toxins occasionally include the antituberculosis drug isoniazid, a urinary antibiotic called nitrofurantoin (Furadantin), methyldopa (Aldomet), for the treatment of high blood pressure, ketoconazole (Nizoral), a potent antifungal, and a host of other pharmaceutical agents and industrial poisons such as carbon tetrachloride, used in cleaning fluid.

Regardless of what caused your cirrhosis, you must still manage its complications. Top priority should be given to restoring normal blood-clotting mechanisms in order to prevent hemorrhage. Early danger signals may be subtle—like persistent oozing from a shaving nick. A substantial hemorrhage may require transfusion of fresh frozen plasma, to restore some of the blood constituents responsible for normal coagulation that are missing or deficient in cirrhosis. Too much *ammonia* in the blood, which results from liver failure, interferes with brain function. Excessive amounts can be reduced by a drug called Lactulose, which is taken by mouth and chemically converts the ammonia in the intestine so that it can be excreted in the stool. Fluid accumulation in the belly and legs requires diuretics. The best combination in liver disease is spironolactone (Aldactone) and furosemide (Lasix).

Until recently, there was no way to slow the rate of scarring in the liver once the process had begun. But *alfa interferon* has recently been shown to arrest or at least retard its progress in some patients with hepatitis B and C.

Finally, the ultimate and best treatment for cirrhosis is exemplified in the case of one of my patients, a sixty-year-old woman with severe liver damage. Her abdomen was filled with fluid, her legs were swollen, she bled repeatedly from her mouth and rectum, and was intermittently confused as toxic levels of ammonia clouded her brain. She had never been a drinker, hadn't been exposed to any liver poison, and could not remember ever being sick with hepatitis. In about one third of all cases we never do find a cause for cirrhosis. However, she did test positive for the hepatitis B antigen—proof of previous infection, which had probably been smoldering over the years.

As she became sicker and sicker, it was suggested that she have an operation to shunt the flow of blood around the obstructing liver. I rejected this solution because I have never known it to be effective, and I was unwilling to add the trauma of futile surgery to her already sorry state. Her only option was a liver transplant, which would not be "wasted" on her since her heart was fine, her kidneys were normal, and she had no other underlying disease. I referred her to the University of Pittsburgh, a great organ transplant center. There her tissues were typed, and a donor liver was found. She received her transplant, and today, four years later, is in excellent health! So, if your cirrhosis is of the galloping variety and seemingly irreversible, and you are still otherwise in good health, remember that liver transplants are not for children alone.

In summary, most patients, and many doctors, are inclined to view cirrhosis as a terminal illness. However, the fact is, there *are* new options that can halt, slow down, or even cure the disease. If you or a loved one are so afflicted, remember the importance of eliminating the contributing cause, which, in most cases, is apt to be the alcohol you've been drinking for years—or any other toxin. If the condition is chronic, and due to old hepatitis B or C, ask about alfa interferon. If that doesn't help, the final option for the lucky few is a liver transplant.

COLD AND WET PALMS AND SOLES:

Except When You're Asleep!

Someone whose palms are cold, clammy, and soggy, like a piece of raw fish, usually has the same problem in the feet and underarms. This condition is called *primary hyperhydrosis.* If you are so afflicted, relax. You're not necessarily neurotic, nervous, or unstable—although anxiety does aggravate the condition—but are suffering from an inherited disorder of the sweat glands. The interesting thing about the abnormal sweating is that it's not at all affected by temperature change; it's not worse when you're hot or better when you're cold. You're always wet *except* when you're asleep, during which time, miraculously, your soggy skin becomes almost completely dry.

The best treatment for hyperhydrosis is *Drysol,* which you can buy at your drugstore without a prescription. Drysol is a particularly good combination of aluminum chloride (the active ingredient in most deodorants) and ethyl alcohol. It works by shrinking the overactive sweat glands. Apply Drysol to the hands at bedtime, and then cover them with a pair of polyethylene gloves. Do this initially for 3 or 4 consecutive nights, then twice a week for as long as necessary. Drysol is usually well tolerated, although some people find it irritating. To treat wet, cold, slimy feet, I suggest the method of Dr. H. G. Hurley, in Upper Darby, Pennsylvania—a solution of 10 percent aqueous gluteraldehyde to which has been added some sodium bicarbonate. You can probably get your pharmacist to make this up for you without a prescription. Apply the solution to the soles of your feet at bedtime 3 or 4 times a week. Although it will also work on the hands, it's likely to turn them light brown.

If these topical treatments fail, all is not lost. You can try a drug called clonidine (Catapres), usually used for the treatment of high blood pressure or drug withdrawal, in a dosage of 0.75 mg, 3 times a day. It occasionally works, too.

CONSTIPATION: When It's Easier to Move Mountains

A patient consulted me the other day because of "constipation" even though there had been no recent change in her bowel habits. Why the sudden concern about constipation? Because, of late, whenever the subject of "regularity" came up at social gatherings (yes, it seems people really do discuss such matters over martinis), everyone present affirmed that *they* enjoyed an evacuation *every* day! Since she was "moving" only every second day, and had been for years, she was worried that she was constipated.

This woman's anxiety is not unusual. Many people believe that daily bowel movements are not only the norm, but absolutely necessary—and they're wrong! How often you "go" depends on your diet, your metabolism, your level of physical activity, and what medications you're taking. It's very much like sex—some engage in it 3 times a day, others 3 times a month, and most of us somewhere in between. You're not really constipated unless you've had a *continuing change* in your usual evacuation pattern. For example, if it was always every single day like clockwork, but for the past week or more it has been every 3 or 4 days, and the stool has become hard and dry, then you are indeed constipated. But that doesn't mean you've got to rush to the nearest drugstore for an over-the-counter laxative like they tell you to in the ads. Give it a little time. Explore some of the following possibilities first. For example, have you been drinking less water lately, and eating too little fiber? Have you suddenly stopped exercising as you used to? Have you become "too busy" to heed the call of nature? Have you started *any* new medication that can slow down bowel function, like narcotics (including those found in cough mixtures), other pain killers, antacid pills, cholesterol-lowering medications like Questran or Colestipol, antidepressants—indeed almost any drug? Such drugs and changes in life-style can reduce the frequency of your bowel movements. In short, try to determine *why* you're constipated instead of looking for a quick laxative fix. If rather than identifying and correcting the underlying cause you just keep taking purgatives, your bowel will eventually become hooked on them and you will have trouble enjoying normal unassisted movements in the future. But if the answer to all these questions is no, and especially if you're not only constipated, but your belly hurts, too, then consult your doctor. Sudden constipation associated with pain *can* be an important danger signal—one that may require looking into.

Here is the best way to deal with constipation that is due to a change in life-style or habits. To get back into the swing of things, drink at least 10 glasses of liquid every day, preferably water; begin a regular exercise program (walking 2 or more miles daily is a good start); eat lots of fiber (whole grain in breads or cereals as well as fruits and vegetables); and drink a glass of prune juice at bedtime. Set aside the same time every day, after a meal if possible, for moving your bowels, regardless of whether or not you have any "feelings." Just sit there and focus on the question at hand. Don't read—that will only distract you. Follow all these recommendations for several weeks, and I'll bet you dollars to Ex-Lax that your constipation will clear up.

If your "functional" constipation (meaning there is no underlying disease causing it) doesn't respond to these measures, you may then "take something." But try not to use *any* laxative for longer than a week or 2 at the very most. Given that caveat, which is the *best* one of a *bad* lot?

Laxatives fall into several categories. I much prefer the *bulk-forming* type, which, even though they do not work as quickly as some of the others, are not nearly as toxic. Such laxatives act by absorbing water *into* the bowel. This extra or "imported" water mixes with the stool and results in a large, soft, more easily propelled and evacuated mass of waste. The best known among these substances is *Metamucil,* an indigestible fiber (hemicellulose) derived from the seeds of the psyllium plant. But be sure to drink a full 8-ounce glass of water with every dose. Unless you do, the Metamucil won't be nearly as effective, and may even cause obstruction of the bowel.

Saline laxatives retain extra water in your bowel—that is, they reduce its reabsorption into the body (the bulk formers actually suck *additional* water into the large bowel or colon). This enhanced "wetness" renders the stool softer and easier to eliminate. The salts required for this laxative effect are phosphates, citrates, magnesium, and sulphates. (Now you know why that familiar white liquid in the blue bottle is called "Milk of Magnesia.") The saline laxatives are best for a fast, no-nonsense, and predictable result, for example, when you need to be cleaned out before a bowel study (X-ray, sigmoidoscopy, colonoscopy) or prior to abdominal surgery, or after you've had an X ray of your colon and the barium has, inexcusably, been allowed to harden like cement in your gut. Even though you can buy these preparations over the counter, I recommend them only in those specific situations and *never* for prolonged use. Avoid them if you have any kidney trouble, high blood pressure, or heart failure, all of which can be aggravated by the additional salt load these laxatives impose.

I'm even less enthusiastic about habitual use of the *bowel stimulant laxatives*—also available without a prescription. Their commercial names are familiar ones—Ex-Lax (whose chief ingredient is phenolphthalein), Dulcolax (bisacodyl), Neoloid (what child hasn't heard of castor oil?), Senokot (contains senna, a plant), aloe, cascara, and others. Dulcolax, which is perhaps the best known, is very effective, comes in oral or suppository forms, and the latter usually works within 15 minutes to an hour. *Never chew Dulcolax* because it contains ingredients that can damage the lining of the mouth and upper intestinal tract on contact. Even the suppository can occasionally cause rectal irritation and cramps (as can any laxative). The senna, aloe, and cascara group, all derived from plants, may color your urine from brown to violet because of a chemical reaction. It's harmless, but a frightening sight. The bowel stimulants are safe and effective *when taken for a day or two now and then,* but their prolonged use can leave you laxative-dependent, and you may end up with a big, sluggish, lazy bowel that regularly *needs* to be whipped into action.

Don't confuse *stool softeners* with laxatives. The former don't cause a movement to happen. They do, however, make it easier to eliminate it without straining. That's important if you've been confined to bed for any length of time, have painful hemorrhoids or an anal fissure, and every hard movement is agonizing. I recommend docusate (Colace), 100 mg, twice a day, but these preparations are pretty much all the same. Remember, however, that *stool softeners won't help constipation.* And don't take them indefinitely, as so many patients do after leaving the hospital. Some of their ingredients are absorbed from the bowel, and no one knows what effect that can have over the years. Another reason to limit the use of stool softeners is that they coat the lining of the bowel and prevent the absorption of vitamins and other important ingredients in your food.

Mineral oil is popular with some older patients who want their bowel "lubricated." I'm not keen on it because, like stool softeners, mineral oil can interfere with the absorption of vitamins. Also, if you take it regularly, you'll leak from the rectum *between* bowel movements. Messy business! Furthermore, if you have trouble swallowing (as older people sometimes do), the oil may be accidentally inhaled or aspirated into the lungs, and cause a very serious form of pneumonia. However, if you're determined to take mineral oil, use the enema route, so that none of it gets into the lungs or interferes with the absorption of nutrients from the bowel. *Never take mineral oil and a stool softener together* because the latter allows the bowel to absorb the oil, which is toxic to the body tissues.

Many persons wrongly believe that *glycerine suppositories* are bland and inert. The fact is, they work by actually irritating the lining of the rectum, causing the reflex that results in the evacuation, usually about half an hour after insertion. Although it's okay to use these suppositories occasionally, I rarely recommend them because my patients have frequently complained of burning and pain in the rectum. Instead, use a plain *tap water enema,* the most effective short-term treatment for occasional constipation. Make your own, or buy it in a convenient, disposable container with a lubricated tip (the Fleet's in!).

A final word of caution: If you've suddenly become constipated, have abdominal cramps, and are literally unable to pass *any* stool or even gas, *do not take any laxative. See your doctor immediately.* You may be suffering from acute intestinal obstruction—total blockage of the bowel from scar tissue (adhesions), previous surgery, a tumor, or *too much* fiber in your diet and not enough water.

CONTRACEPTION: When You're Neither Ready Nor Willing

Pregnancy is not a disease, and contraception is not a treatment, so why include them in this book? Because, in my view, contraception *is* tantamount to treatment when an unwanted pregnancy is a potential threat to a woman's life, or carries with it harmful psychological, social, or economic consequences, and its termination by abortion contravenes the mother's religious or moral principles.

Although contraceptives do vary somewhat in their effectiveness, you will get the best result with whichever one you select if you use it correctly. For example, a condom will prevent pregnancy only if you make sure it doesn't leak or spill over; a diaphragm won't do you much good if it's the wrong size or you forget to add the spermicide; oral contraceptives *must* be taken exactly as directed.

The "Pill"

Assuming that you follow all the directions, the safest, most effective (almost 100 percent), and most convenient form of contraception is the Pill. It *works basically by preventing ovulation.* No egg—no baby! However, it also alters the quality of the mucus in the cervix (so that it is more difficult for the sperm to gain entry to the uterus), and affects the lining of the uterus in such a way that even if an egg is fertilized, it will not "take hold."

There are some forty brands of "Pill" to choose from, and they all contain estrogen and progestin. But the preparations vary considerably in terms of how much of each of these hormones is present. The Pill you select should have no more than 35 mcgm of estrogen and 0.4 or 0.5 mg of the progestin (Ortho Novum, Ovcon 35). Most important, however, is to take it *exactly* as prescribed.

Safe and effective as it is, do *not* use the Pill if:

- You have a history of *blood clots* in the past. (Estrogens promote blood coagulation and embolism in women who are so disposed.)
- You are over the age of thirty-five *and* are a smoker.

- You've had a stroke, angina pectoris, or coronary artery disease.
- Your *cholesterol level* is greater than 250 mg (progestin raises cholesterol levels).
- You have any kind of *liver trouble.*
- You've had breast cancer (a previous malignancy of the ovaries or uterus would have required their removal making contraception unnecessary).
- You have abnormal, undiagnosed *vaginal bleeding.*

These are the *absolute contraindications*—that is, circumstances under which you should *virtually never* take the Pill. But there are situations when, although it is not forbidden, you are better off avoiding it, for example, if you're diabetic, have gallbladder disease, suffer from migraine headaches, or are confined to bed for whatever reason (and so are at increased risk for phlebitis and embolism). Also, if you're planning to have an operation in the next 4 weeks, it's a good idea to switch from the Pill to another contraceptive because surgery and estrogens both promote blood clotting.

This list of "verbotens" may seem very long, but it actually does not apply to as many women as you might think, and the vast majority *can* and *should* use the Pill. Remember, however, that even when they're safe for you, oral contraceptives can cause breast tenderness, vaginal bleeding, nausea, and other side effects.

Many women refuse the Pill because they're afraid it will lead to cancer of the breast and the uterus. The truth is that the combination of estrogen and progestin I recommend does not increase the risk of uterine and ovarian malignancy. But the jury is still out with respect to breast cancer. Frankly, I am not convinced by any of the data I have seen so far that the Pill, in fact, raises that risk. But just to be sure, if you're now on the Pill, or have been in the past, you should have regular breast exams and mammograms after age forty (every 2 years between forty and fifty, annually thereafter), especially if there is also a history of breast cancer in your family.

The Diaphragm

The diaphragm, whose failure rate is slightly higher, is a suitable and popular alternative to the Pill. It should not, however, be the very first contraceptive you use. Nor do I recommend it if you've had a baby in the last 3 or 4 months (because the size of the vagina is so variable, especially during breast feeding, that the diaphragm cannot be reliably fitted), or if you suffer recurrent attacks of cystitis or urinary tract infection, which may be aggravated by the diaphragm.

If you decide on a diaphragm, remember that you can't just go into a drugstore, buy one, and insert it yourself. It must be fitted by your gynecologist and checked once a year to make sure that the size is still right and that there are no holes or defects in it. Because that's something of which you cannot be certain, always add spermicide. After your first fitting, ask your partner to use a condom for a couple of weeks until you're adept at inserting and removing your new diaphragm.

The Cervical Cap

The cervical cap is a variation on the theme of the diaphragm. It is made of rubber but is smaller, and fits over the cervix like a thimble. Although it has the same small failure rate, one advantage of the cap over the diaphragm is that you can leave it in place for 24 hours without adding additional spermicide, something that must be done with the diaphragm every time you have sex.

The Vaginal Contraceptive Sponge

The vaginal contraceptive sponge, which contains the spermicide nonoxynol 9 (Today), does not need to be fitted. You simply wet it, insert it into the vagina, and it's good for any amount of sex for the next 6 to 12 hours. Although it's less messy than a diaphragm with spermicide, it has a higher failure rate than the cap, the diaphragm, or the Pill. So don't depend on it if you've had several children already, and another one would "break the bank."

The Intrauterine Device (IUD)

Intrauterine devices used to be much more popular than they are today. Most of them have been withdrawn from the market largely because of the litigation and problems associated with the Dalkon Shield. Only two IUDs—both T-shaped—are available in the United States at this time. My patients prefer ParaGard, which is impregnated with copper and can be left in place for about 6 years. It is virtually as effective as the Pill and costs about $300 (that includes the gynecologist's bill for inserting it). The other IUD, called Progestasert, contains progesterone and requires changing every year.

If you are likely to want a baby in the near future, there are several reasons not to use an IUD. Women who do have a higher incidence of ectopic pregnancies. More important, unlike the diaphragm, cervical cap, and sponge, the IUD is not just a mechanical barrier. It also alters the lining of the uterus, called the endometrium, leaving it inhospitable to any sperm that might make it inside. Since the endometrium must be intact in order to nurture and sustain a pregnancy, it's the one area with which you don't want to tamper—and the IUD does. Finally, you're probably better off not using an IUD if you have a cardiac valve problem, since it can render the heart vulnerable to infection.

Condoms

The condom is the best technique for random sexual encounters because, in addition to being a good contraceptive, it also protects against most sexually transmitted diseases, including AIDS. But all condoms are not equally effective in that regard. Those made of lamb intestine permit penetration by the HIV virus; the rubber/latex condoms do not. (If you are particularly worried about AIDS, buy a rubber condom coated with nonoxynol 9, which appears to offer even greater resistance to such infection.) The condom has no side effects, and has a track record somewhere between the diaphragm, cap, and sponge on the one hand, and the Pill and IUD on the other.

Norplant

The newest birth control option is a long-acting hormone, levonorgestrel (Norplant). It is sealed in a plastic cylindrical capsule less than an inch long, which is implanted through a tiny incision under the skin on the inside of the upper arm. The drug is continuously released from pellets, and provides a level of protection equal to that of the Pill, in other words, virtually 100 percent, for up to 5 years. The capsule can be removed at any time. But even though future pregnancies are not compromised or endangered, I recommend this contraceptive route only to women who have decided that they are not likely to wish to become pregnant again sometime in the future. My reason for doing so is because of Norplant's potential side effects—irregular periods with heavy bleeding, headache, acne, weight gain, and unwanted hair growth or loss. Although these do not occur in all women, they are frequent enough for you to consider when you're making a decision about going that route.

Depo-Provera

Look for the FDA decision concerning Depo-Provera. Though not yet approved in this country, this hormone has been widely used virtually everywhere else in the world. A single injection every 3 months confers contraceptive protection equal to that of the Pill. I understand that the reason for the delay in approval is the fact that it can cause irregular bleeding in humans and breast cancer in dogs.

The Rhythm Method

The religious persuasion of many women permits them only the technique of periodic abstention timed with the ovulation cycle to avoid pregnancy—the method with the highest failure rate. If that's the only choice available to you, ask your gynecologist for *detailed* instructions on how to test the cervical mucus and to carefully explain the rhythm method to you.

Tubal Ligation and Vasectomy

In tubal ligation the fallopian tubes that connect the ovaries to the uterus are tied off. Vasectomy in men involves tying off the ducts that deliver the sperm from the testes to the urethra, whence they are ejaculated. These two surgical techniques provide permanent contraception. Should you change your mind, these structures can be untied, but at additional expense and without complete certainty of success. So don't make the decision to have them capriciously. Another reason to think twice is a recent observation that men who have had vasectomies may have a higher incidence of prostate cancer later in life.

COUGH: How and When to Stop It

A cough, though annoying, is beneficial when it rids the air passages of unwanted "gunk." But when it's "dry" so that you are just hacking away and not spitting anything up, it should be suppressed. The best way to do that depends on what's causing the cough. Here are some of the conditions that will give you a tickle in the throat, and the measures that I have found to be most effective for stopping it.

- *When you have a cold and a stuffy, runny nose,* the accompanying cough is usually due to mucus dripping down into the back of the throat. A *decongestant* that shrinks the swollen blood vessels in the nose and opens the nasal passages will prevent this postnasal drip, and is the best treatment for that type of cough. The decongestants I recommend are phenylephrine, pseudoephedrine, or phenylpropanolamine. Most "cold" formulas contain one of these preparations in combination with other substances, but I prefer you take it alone. All decongestants cause a "rebound" phenomenon—that is, when you discontinue them after a few days, the congestion of the blood vessels in the nose recurs, with a vengeance! Then you're stuffier than ever. This rebound is less likely when the drug is taken orally, which is why I recommend the tablets or liquid rather than the nose drops or sprays. But whichever route you choose, do not take any decongestant more often than 3 or 4 times a day or longer than 3 or 4 days. And incidentally, stay away from alcohol when your nose is stuffed because it worsens the congestion.

Decongestants are safe when taken *exactly* as prescribed, but avoid them if you have high blood pressure, thyroid disease, or a heart problem. Sharing your sprays and droppers with other family members means also sharing whatever infection you have in your nose. And don't give a decongestant to a child under six years of age unless the pediatrician assures you that it's okay.

Most cold preparations contain an antihistamine, in addition to a decongestant, ostensibly to help "dry" the nasal passages. I do not believe antihistamines have any such effect. In fact, I advise you avoid them because of their potential side effects—dry mouth, drowsiness, and, in older men, interference with normal voiding. But you'll have a hard time finding an antihistamine-free cold remedy. Two that fit the bill are Ornex and Endecon. And while you're checking the labels, also look for caffeine content and avoid that, too, if you can. Caffeine does nothing more than counteract the drowsiness caused by the antihistamine that shouldn't be there in the first place!

• *If you have a cold and are coughing, but your nose is not stuffy,* then the cough is probably due to viral irritation and inflammation lower down in the respiratory passages. If it's a wet cough, and you're spitting up globs of yellow or green mucus, *do not suppress it*. Get rid of as much of this "junk" as you can. But after the cough becomes dry—that is, nonproductive—then take something for it. At that point it's only irritating your airways. It's a vicious cycle—coughing irritates the airways, which leads to more coughing, and so on. Although *codeine* is the most effective cough suppressant, I prefer *dextromethorphan,* which is present in many over-the-counter preparations because (a) it's not a narcotic, (b) it's *almost* as effective as codeine and usually all you'll need, (c) it's less constipating than codeine, (d) you can get it without a prescription, (e) it is not habituating, and (f) it's safe. If, however, dextromethorphan doesn't work as well as you'd like, then switch to codeine.

The maximum daily adult dose of codeine for this purpose is 30 mg, 4 times a day; children under six should not take more than 5 mg, 4 times a day. The upper limit for dextromethorphan is 20 mg every 4 to 6 hours for adults and much less than that—5 mg, 4 times a day—for children under six. How long you will require either drug depends on what's causing the cough. See your doctor if it hangs on beyond 2 weeks.

Regardless of which antitussive you choose—codeine or dextromethorphan—take it "straight" in tablet form. There's no need to gag on a syrup that often tastes terrible, may depress an already impaired appetite, and usually contains sugar, which you want to avoid if you are diabetic (although there are sugar-free cough syrups). The point is, none of the added ingredients in the syrup is going to help your cough.

There's a big market out there for something called "*expectorants.*" They're supposed to "loosen" your cough so that you can spit out the mucus in your airways. There are several different commercially available preparations including potassium iodide, hypertonic saline, and guaifenesin. In my view, you're probably wasting your time, money, and effort on any of these preparations. With the exception of guaifenesin, which *may* have some liquefying effect in very large doses, I have never found these products to work.

What about a humidifier? I remember using, as a child, the hot, humid kind—and loving it. Today, some doctors believe it may actually prolong symptoms. You're probably better off with the cold mist rather than the steam.

If you're still coughing after 2 or 3 weeks, arrange for a thorough physical exam and a chest X ray just to rule out the possibility of heart failure, lingering pneumonia, an allergic disorder, cancer, AIDS, or some exotic fungal disease. But chances are your doctor will find that your chronic cough is the result of one or more of the following:

- Cigarette smoking
- Postnasal drip
- Bronchitis
- Asthma
- Reflux from your stomach up into the food pipe
- Side effect from blood pressure medication

Here's how to deal with each of the above:

• *Smoker's cough* is due to chronic irritation of the airways by the inhaled tobacco fumes, and there is no medication that can prevent it. Don't depend on codeine for relief—you may end up a constipated codeine junkie as well as a nicotine addict. No matter how long you've had the cigarette habit, stop smoking *now* and you can be sure that your cough will either clear up or improve substantially in a few weeks. You'll also feel better, your food will taste better, your shortness of breath will clear up, and you'll live longer.

• *Chronic postnasal drip* (not the kind that clears up a few days after a cold) is due either to *sinus infection* or to *allergy.* In the latter case, avoid whatever it is to which you're allergic, and use an antihistamine. I prefer a 60 mg tablet of Seldane (terfenadine) twice a day for daytime use. It's less likely to make you drowsy at the wheel of your car or on the job. At night, take any antihistamine; they all promote sleep. Elderly men with large prostates should be careful using antihistamines since they can make urination more difficult. If antihistamines do not stop the drip and the cough, then a steroidal nasal spray (*Vancenase, Nasalide*), 2 puffs, 3 or 4 times a day for a week or two, almost certainly will—by reducing the swelling, inflammation, and allergic response of the nasal passages.

• Postnasal drip due to *chronic sinusitis* makes you cough, and leaves you with a feeling of fullness in your face as well as a headache. Infected sinuses are hard to treat because they are closed spaces surrounded by bone, which most antibiotics have a tough time penetrating. Amoxicillin, 500 mg, 3 times a day for at least 2 weeks (but not if you're allergic to penicillin), is particularly effective because it is active against most of the bacteria buried within the sinuses. And don't quit before the full 2 weeks. It takes that long for the amoxicillin to do the job. A nasal decongestant is also helpful here because it allows the sinuses to drain.

• *Asthmatics* do a lot of coughing, too. Control the asthma and the cough will stop. The best regimen consists of a combination of bronchodilators (by inhalation) and *steroid* inhalers (not quite the same as steroid nasal sprays). For a more detailed discussion, see pages 30–35.

• You may experience a chronic cough if food and/or secretions from the stomach regurgitate up into your *esophagus* (food pipe), an action known as *reflux.* Unless you and your doctor are aware of this possibility, you'll be referred to lung specialists, allergists, and ear, nose, and throat physicians, all of whom are likely to focus on a respiratory cause for your cough when the trouble really lies in the upper intestinal tract. The correct diagnosis can be confirmed with an upper GI series (X ray). Here's how best to deal with such reflux: Do not eat for 3 hours before going to bed (so there's less in the stomach that can regurgitate when you lie down), eat less at each sitting and more frequently, use any one of the many available antacids 3 or 4 times a day *between meals,* and elevate the head of your bed with 6- to 8-inch blocks (it's more difficult for reflux to occur when you're sleeping "high"). If these measures don't stop the cough, try one Prilosec (omeprazole) tablet a day—it comes in 20 mg strength and requires a prescription. This drug will often work wonders in such cases.

• The key to controlling the cough of *chronic bronchitis* is to stop smoking. In addition, during flare-ups, the right antibiotic will eliminate the infection in the bronchial tubes. But no antibiotic will stop the cough for long if you continue to smoke. The best cough therapy in chronic bronchitis is Atrovent, available only in aerosol form. Take 1 or 2 puffs, 3 or 4 times a day. It's safe and effective (provided you do not have glaucoma). I prefer it to codeine for long-term use.

• Some blood pressure medications cause a persistent cough. A woman consulted me some 2 or 3 years ago because of a chronic dry cough. She was not a smoker, and aside from mild high blood pressure, which was being nicely controlled with medication, she was in excellent health. She had not lost weight, and she was not aware of any allergies. Physical examination revealed nothing unusual. She had no fever, her lungs were clear, she didn't have a postnasal drip, her blood count was normal, and her chest X ray was completely clear—yet she was coughing. Some medical detective work unearthed the fact that she was taking 10 mg of enalapril (Vasotec) for her blood pressure—and that drug proved to be the villain. Within days after switching her to another blood pressure medication, her cough stopped. Enalapril is one of a very important class of drugs introduced for the control of high blood pressure and heart failure a few years ago—the *ACE* (angiotensin converting enzyme) *inhibitors*. They're marketed here as Capoten, Vasotec, Zestril, and Prinivil. More are on the way. They all have one unwanted side effect in some patients: a persistent dry cough. If you're taking any of these preparations, and there's no other explanation for your cough, it will probably disappear in a few days after you stop the drug.

DIABETES: How To Live Longer and Better (But Not Sweeter)

There are many more diabetics these days than there used to be (15 million Americans alone, and half of them don't know it), not only because they're living longer, but because the incidence of the disorder itself is increasing.

Most nondiabetics don't think this condition poses much of a problem. As far as they're concerned, all one has to do is keep away from sweets, and if that's not enough, there are pills or insulin to lower the high blood sugar. But diabetics know how complicated life can become, and are full of questions about how to manage their life-style. I have tried to anticipate and answer those questions in the following pages.

One key fact to appreciate is that diabetes is not a single disease. Although all diabetics have abnormally high blood sugar, the mechanisms responsible for it are not necessarily one and the same. For example, diabetes appearing in childhood or adolescence always requires insulin, is often difficult to regulate, and has a significant impact on the quality and duration of life. By contrast, the adult-onset type usually does not need insulin, sugar levels are generally stable, and the risk to life is much smaller.

The pancreas, which lies deep in the belly behind the stomach, virtually on your backbone, makes the insulin, which regulates sugar metabolism. In *juvenile* or *insulin-dependent diabetes,* there is very little insulin around because the special cells that produce it within the pancreas have been destroyed by the body. This is a good example of an autoimmune disease in which a "normal" constituent is perceived by the body to be an "enemy" and so "killed." We're not sure why this happens in some children and not in others. It has been suggested that a viral infection early in life is the trigger for this bizarre response. Regardless of the underlying cause, these diabetic youngsters have virtually no insulin. Now, as you know, the sugar you eat is absorbed from the stomach, enters the bloodstream, which carries it to every tissue in the body where it provides the energy necessary to maintain life. But in order to do that, the sugar must first penetrate the cells and undergo various chemical transformations, a process that requires insulin. If you have too little of this hormone, the sugar that can't get into the tissues just floats in the blood until it is excreted in the urine—a total waste. Since there isn't enough sugar available as a source of energy, the body falls back on its protein and fat reserves. Unless the missing insulin is administered, this alternative solution, the continuing utilization of fat and protein instead of sugar to fuel the body, results in a long chain of consequences and complications whose most obvious manifestation is coma.

By contrast, in the *non-insulin-dependent* or adult-onset form of diabetes, which usually develops after age forty and whose incidence peaks between forty-five and sixty-five years, there is no shortage of insulin. But in order for the insulin to do its job, the cells on which it acts must be able to "accept" it, and that requires "insulin receptors." Adult diabetics, especially fat ones, suffer from a deficiency of these receptors. So we don't usually give these people insulin since they have enough of their own. Instead, we prescribe a diet, especially if they're obese, and have them lose weight, all of which results in an increased number of insulin receptors and thus a lower blood sugar.

This knowledge about how insulin works—the juvenile versus the adult form of diabetes, insulin receptors, the autoimmune concept—has all come to light only in the last decade or so. Leafing through a medical book published in the early 1900s, I found the *symptoms* of diabetes—thirst, frequent urination, weight loss, increased appetite, itching, and coma—accurately described. But although high blood sugar was recognized as the cause, insulin was not destined to be discovered until the 1920s. In this same text, the author stated that most children with diabetes live only 6 months to 4 years after the diagnosis is made. Death in those days came in the form of "exhaustion" or coma. Treatment consisted of an absolutely rigid diet without even a trace of sugar or starch—no bread, milk, potatoes, carrots, and not a single fruit. No wonder these kids died so soon!

After the discovery of insulin (which was rewarded by a Nobel prize), death from coma declined dramatically. But as more and more diabetic children reached adult life, we began to realize that diabetes damages blood vessels everywhere—in the brain, the heart, the eyes, the legs, and the kidneys. That's why this disease is still a major killer and crippler, the seventh leading cause of death in this country. Although children are hardest hit, the overall lifespan of diabetics of any age is only two thirds that of the general population. But you can improve on those statistics if you know how best to deal with your disease.

The most frequent questions my diabetic patients ask me are, "Must I keep my sugar level rigidly within normal limits at all times? How dangerous is it to 'cheat' now and then?" I tell them that most doctors believe that the closer to normal you maintain your sugar over the years, the less likely you are to suffer the deadly complications of diabetes. In my view, that's especially important for the young insulin-dependent patient. I don't feel as strongly about such rigid control in the adult, where other factors like cigarette smoking, hypertension, and weight control are at least as important as the sugar level. So my advice to maturity-onset diabetics is to try to maintain a blood sugar level that keeps them symptom-free rather than have them focus on target numbers. Diet, exercise, oral medication, or insulin—the bases of diabetic treatment—should be so balanced that excessive thirst, the need to get up frequently at night to void, weight loss, fatigue, and other evidences of a high blood sugar are minimized. But equally important, especially in the elderly, is the avoidance of a blood sugar that's *too low.* Such hypoglycemia, especially in someone with heart or vascular disease, is much more dangerous than is a modestly elevated sugar. Not only does a sharp drop in sugar increase the risk of heart attacks and strokes, it's hazardous for other reasons. For example, consider the impact of an attack of low blood sugar that impairs your thinking, or causes a loss of consciousness while you're performing any act that requires judgment, such as driving a car, or even walking across the street.

Bearing all this in mind, I advise my *diabetic* patients to strive for a fasting level (blood is drawn after you've had nothing to eat or drink since the previous midnight) of less than 140. (It is less than 120 in nondiabetics.) The reading 2 hours *after* you've eaten should not be greater than 50 percent above the fasting level. But if these goals result in attacks of hypoglycemia, then you should settle for higher readings. Remember, better too high than too low!

Not so many years ago there was no way a diabetic could really know how well his or her sugar levels were being managed short of going to the doctor for the necessary blood test; it's a different ballgame now. There are several commercially available glucose measuring units that are very simple to use, and that provide an immediate accurate sugar reading. I have the most experience with Accu-Check III, which I highly recommend. It's priced at about $150, and can be purchased at most drugstores or surgical supply houses.

The bases of diabetic treatment are diet, medication, and exercise. Let's consider diet first. The three key factors to bear in mind are (a) *the total number of calories* you eat, (b) the *fat*, *protein*, and *carbohydrate* content of those calories, and (c) the *time relationship* between eating, taking any sugar-lowering pills or insulin, and exercising.

With respect to the number of calories consumed, remember that nutrition is very important for the growth and development of the diabetic child. Calories should not be restricted, especially if the youngster is already underweight, as many are. Such kids should be given as much insulin as is necessary for them to be able to eat a body-building diet and at the same time maintain a close-to-normal blood sugar. By contrast, overweight, middle-aged diabetics should restrict their caloric intake to whatever extent is necessary to lose weight. But don't mess with those liquid diets or powdered food substitutes. You're much better off consulting a dietician for guidance in putting together a menu that you can follow over the long-term. If, however, you cannot resist the lure of the ads for the quick-fix weight-loss programs, then when you start one, be sure to decrease the dosage of the oral diabetic drugs or insulin you are taking. You run the risk of hypoglycemia if you don't.

The diabetic diet should be free of all *simple* sugars like sucrose. That means no candies, cakes, frostings, and other delicious foods. But, *complex* sugars of which there are many—pastas, fruits, beans, and nuts—formerly forbidden, are now considered permissible. In fact, such *carbohydrates should make up 50 to 60 percent of your total calories.* Any soluble fiber like oat bran is good, too, because it lowers both your sugar *and* your cholesterol. I also advise my diabetic patients to supplement their diet with a 200 mg a day tablet of chromium picolinate, a supplement available at health food stores and pharmacies. It may make it easier to control sugar levels. If you miss the sweet taste in your diet, you can add one of the artificial products like aspartame (which the FDA has certified as "safe"). But remember that like any other chemical, sweeteners can cause side effects, especially when used in large amounts.

Your fat intake should be less than 30 percent of the total calories. There are three kinds of fat—*saturated* fats (found in animal foods), *polyunsaturated* fats (present in most vegetable oils), and *monounsaturated* (such as olive oil). Saturated fats should constitute *less* than 10 percent of your total fat intake. Since diabetics are more vulnerable to arteriosclerosis, also limit cholesterol consumption to no more than 300 mg per day.

Having assigned approximately 50 percent of the caloric intake to carbohydrates and no more than 30 percent to fat, you have approximately 20 percent left for protein—basically meat, poultry, and fish. I suggest you go heavy on the fish and poultry, and light on the red meat. You're also better off not drinking, but if you find that easier said than done, *limit your alcohol* to two glasses of wine a day (or a bottle of beer, or one cocktail).

Exercise is more important for diabetics than it is for most other people. But remember that it lowers blood sugar, so if you are on insulin or an oral diabetic agent, *always eat something extra before working out vigorously.* If you plan to exercise for prolonged periods, take some food along with you and snack at least every half hour, and either reduce the dosage of your medication or increase your food intake before beginning the workout.

What are your options if you're a "non-insulin-dependent" diabetic, and despite a good diet and regular exercise, your sugar level simply won't drop below a fasting value of 180? In such cases, I prescribe either insulin or sugar-lowering pills, but rarely both. The pills make your tissues less resistant to insulin, and they also "drive" the cells in the pancreas to produce more of it. These medications have been available for many years, but I prefer the newer derivatives, specifically glyburide (Diabeta, Micronase). They have fewer side effects than the older drugs, and are more potent. By the same token, however, they are more likely to cause hypoglycemia, so use them as cautiously as you would insulin.

Insulin, which is always given by injection, is derived from beef, pork, mixtures thereof, and more recently, thanks to recombinant techniques, human sources. I prescribe human insulin, because it causes fewer allergic reactions. But if your disease is well controlled on one of the other forms, there's no reason to switch. If, however, the beef or pork shots are painful or are causing lumpy fat formation at the site of injection, then by all means change over to one of the human brands.

Here's some practical advice:

• *Always wear a bracelet* (*Medic Alert*) *indicating that you are diabetic,* especially if you're prone to attacks of hypoglycemia (your doctor can tell you where to buy it, or if he can't, write to the American Diabetes Association for a list of dealers). A diabetic who has lapsed into coma because of very high or very low sugar may be mistaken for a drunk or a drug abuser.

• *Always carry some form of sugar in your purse or pocket,* just in case. If you feel an attack of low blood sugar coming on (because you're nervous, sweating, trembling, or can't think straight), take orange juice or candy immediately. Remember, too, that a brain temporarily deprived of sugar doesn't function normally immediately after apparent recovery. So if you've been hypoglycemic, wait for at least an hour after taking sugar before you resume such activities as driving a car.

In summary, the keys to the successful management of diabetes are insulin, diet, and exercise for the young, and diet, weight control, and exercise for most adults. Oral medication and insulin may be required in some adults if their sugar levels cannot be reduced any other way. Although a normal blood sugar level is desirable, it's not worth it if you keep getting attacks of low blood sugar. Hypoglycemia is more hazardous than sugar levels that are somewhat above normal.

DIARRHEA: When Running Isn't Good for Your Health

Diarrhea—the passage of unformed, loose, or watery stools—may be mild, (only one or two movements a day), or severe (virtually constant, every hour on the hour—like the news); it may begin suddenly, run its course for a few days, then stop, or it can be chronic and plague you throughout your life. Frequent movements that are solid are not considered to be diarrhea.

Diarrhea that comes on abruptly and ends in a day or two is usually caused by tainted food or a virus. The first, best, and only thing to do in such cases is to replace the fluid lost during those 24 or 36 hours. *Do not treat diarrhea by starving yourself.* Take lots of broth, bouillon, or chicken soup, and add extra salt (unless you happen to have kidney or heart trouble). Fruit juice is good, too, except for apple juice, which sometimes worsens diarrhea in children. That's probably due to its sorbitol (a sugar) content. But any other sugar will help. Avoid coffee, which is a bowel irritant. There are also specially constituted commercial preparations available to replace the fluid lost in the diarrhea, but homemade drinks are usually perfectly adequate.

It's a great temptation to try to take something to stop diarrhea as soon as it develops. Hold off for at least 12 hours because diarrhea is the body's way of ridding itself of whatever toxin, virus, or bug is causing it. Give nature a chance! But if the symptoms continue after this initial "cleansing" period, take 2 of the 2 mg Imodium (loperamide HCL) tablets, 4 times a day, or after each loose movement (but no more than 8 tablets or 16 mg per day). You can buy it over-the-counter without a prescription.

Traveler's diarrhea, which also begins abruptly, is managed somewhat differently than an acute viral infection. The best way to avoid "turista," wherever you are, is to be careful about what you eat and drink. "Boil it, peel it, cook it, or forget it" should apply to everything on the menu of even the most sophisticated restaurant in most Latin American, Asian, or African countries—and whatever you eat should be prepared just before it's served. Cooking it ahead and then having it sit around breeds bacteria.

Many travelers who scrupulously follow all the rules still come down with diarrhea. So if you're going somewhere "infectious," you should consider taking something prophylactically. I prefer sulfamethoxazole (Bactrim

DS, Septra DS—identical products with different trade names). Two the day before you leave for your trip, then daily while you're abroad and continued for a couple of days after you return, will greatly reduce the risk of diarrhea (provided, of course, that you remain careful about what you eat and drink). Bactrim has very few side effects, but don't use it if you're sensitive to sulfa. For persons who are, I advise 1 tablet of Noroxin (norfloxacin) a day (unless you're under twelve or pregnant).

Are you one of those people who "hates to take pills," who refuses to be bothered with Bactrim or any other prophylactic medication, who is sure that just being careful is enough? You may be in for a surprise, because despite all your precautions, you may wake up one morning with loose stools and abdominal cramps. What should you do now? If you have no fever, let nature run its course for a day or two. Chances are your symptoms will clear up within 48 hours. While you're waiting, avoid solids even if you're hungry, and drink *lots* of liquids to replace the lost fluids, just as you would for any kind of diarrhea. But if you're on your honeymoon or are an athlete competing in international games, or involved in important business or diplomatic transactions, it's hard to fulfill your obligations from "the throne." In that case, 15 mg or 30 mg of codeine every 4 hours will usually slow things down. (Codeine is a wonderful drug—for pain, cough, or diarrhea.) Three or 4 Imodium (loperamide) or Lomotil (diphenoxylate) a day are effective, too. (You need a prescription for Lomotil but not for Imodium.) All these agents work by reducing the contractions of the bowel. Kaolin and pectin (a commercially available brand, Kaopectate, combines the two) improve the diarrhea by causing the bowel to absorb large amounts of water. Kaopectate is not usually strong enough for adults except when the diarrhea is mild, but it is effective in children.

You should see a doctor promptly if your diarrhea is accompanied by abdominal pain, severe chills, or blood. This is especially important if you have anything artificial implanted in your body—a heart valve, or a new hip or knee—which is vulnerable to serious infection.

If you're still "running" after 7 days, you may have been infected by an organism other than the relatively harmless one responsible for most cases of traveler's diarrhea. It is important to identify that bug because whatever it turns out to be—an amoeba (see page 14), some other parasite, or a dysentery organism—will require *specific* medication to eradicate it. Imodium, Lomotil, codeine, or Kaopectate may alleviate the diarrhea, but will not cure an infection.

Diarrhea that is *chronic* is usually due to some underlying disorder. In my own practice, I have found lactose intolerance to be responsible for most cases in otherwise healthy individuals. These persons lack an enzyme called lactase, needed to digest lactose, which is present mainly in milk and milk products. Undigested lactose produces flatulence (gas, bloating, cramps) and diarrhea. Other common causes of chronic diarrhea include:

- An irritable (nervous) bowel
- Inflammatory bowel disease (Crohn's, ulcerative colitis)
- Bowel tumors
- Laxative abuse
- Antibiotics
- Surgical removal of a large portion of the gut
- Radiation treatment for a cancer in the area of the lower abdomen (prostate, urinary, or bladder cancer)
- Chronic stress
- Medication (antacids for hyperacidity, quinidine for control of a cardiac rhythm disturbance).

If you have chronic diarrhea stemming from any of the above, the obvious best treatment is to correct the underlying cause wherever possible. You will find the specifics of that therapy in the appropriate sections of this book for each disorder.

A final word of caution: It's best to ask your doctor what to do about *any* medication he has given you if you come down with diarrhea that lasts more than a day or two. That's because when you have diarrhea, your gut is hyperactive, and propels whatever you take by mouth very quickly through the bowel and out of the body. So any oral medication you require for some other problem may not remain in the gut long enough to be absorbed and do its job. That can have important consequences. For example, if you develop severe diarrhea while you're on the Pill, you may not be adequately protected. If you're on heart or blood pressure drugs, their dosage may need to be adjusted.

EAR INFECTIONS IN CHILDREN: When No One Sleeps!

Middle ear infection (otitis media) is the second most frequent reason, after the common cold, for calling the pediatrician. What young couple hasn't been plagued by their feverish child crying nonstop with pain throughout the night in the first years of life? If the infant is too young to tell you what's wrong, he or she will reveal the diagnosis by tugging at the affected ear.

Middle ear involvement most frequently follows some upper respiratory infection, usually a common cold. Before the era of antibiotics, the symptoms would persist for about 2 weeks, after which the infection cleared up spontaneously. However, these days the problem can usually be solved within 72 hours simply by giving the youngster the right antibiotic—amoxicillin, 40 mg for every kilo (2.2 pounds) of weight in 3 divided doses per day. One of the cephalosporin group of antibiotics, such as Keflex, is also effective. In cases of penicillin allergy, use Bactrim DS or erythromycin, again in doses depending on the weight and age of the child. Acetaminophen (Tylenol) will usually control the pain and fever. In children old enough to manage one, a nasal spray like Afrin may also help.

If antibiotics don't work and the ear pain and fever persist, there may be too much fluid in the middle ear, which a doctor can drain by inserting a small tube through a little nick in the drum. Always have your child reexamined one month after a middle ear infection just to make sure there is no fluid left in the ear, and that hearing has not been affected.

ENDOMETRIOSIS: For Women Only!

Many women suffer from *endometriosis,* a condition that causes severe pelvic and abdominal pain or cramping. Symptoms are usually related to the menstrual cycle, beginning a few days before the onset of the period and continuing until it is over, but may occur also during intercourse, urination, or bowel movements. The specific complaints vary from female to female depending on where the endometriosis is located. To understand why and when these symptoms occur, you need to know what endometriosis is.

The lining of the uterus, called the *endometrium,* is composed of special cells that respond to the cyclical hormonal changes that occur in women. This tissue is first enriched with blood so that should conception occur, the fertilized egg will find a nutritious environment in which to settle down and eventually develop into a baby. If the egg never meets a sperm, there is no purpose in maintaining this rich lining, and so it is shed, a phenomenon we call *menstruation.* This process within the uterus, in which the lining is first engorged with blood and later denuded, is repeated every month until menopause. It stops at that time because the hormones responsible for it are no longer produced in the same amounts. For some unknown reason, in women with endometriosis, islands of this normal lining tissue have broken away from the uterus and traveled to other areas in the pelvis—the ovaries, the vaginal vault, the fallopian tubes, and occasionally even to the bowel wall. And although these peripatetic cells have no business being in such foreign locations, they act as if they were still back home in the uterus—swelling, shrinking, and bleeding in response to the monthly hormonal changes. Imagine the pain that a patch of such tissue can cause on an ovary or on the bowel. That's what endometriosis is all about.

There is no cure for endometriosis. Treatment consists basically of making the patient more comfortable. One way to do that is to stop your periods—by pregnancy, for example, but that only works for 9 or 10 months! (If you do become pregnant, you can extend this pain-free interval by breast feeding your infant, since that also results in hormonal changes that suppress stimulation of the wandering endometrial fragments.)

An effective medication for endometriosis is a hormone preparation called nafarelin (Synarel) taken as a nasal spray twice a day for 6 months. But it stops your periods as it controls your symptoms. And, since it works by suppressing estrogen formation, Synarel may give you hot flashes, diminish your sexual appetite, and increase loss of calcium from the bone—exactly what happens in menopause. So take calcium supplements along with it. *Do not use Synarel if you are pregnant or breast feeding, or if the diagnosis of endometriosis is not absolutely certain.*

If for any reason you can't use Synarel, then birth control pills or a drug called Danazol may also suppress endometriosis. If you're lucky and have only a small patch of endometriosis in the pelvis, it can be vaporized with a laser via a laparoscope (an instrument through which the interior of the pelvis can be viewed). Unfortunately, bowel lesions are not similarly accessible.

If the pain of endometriosis is intense and fails to respond to medical management, you may need an operation to remove the tissue that's causing all the trouble. But such surgery is not predictably successful, especially if the islands of tissue are widespread, and should only be done as a last resort.

FEVER: What? No Aspirin?

Fever is a symptom, not a disease. It's also an important natural defense mechanism, and a little can actually be good for you. Many infectious organisms can't "take the heat"; they don't thrive or multiply nearly as rapidly when the body temperature is up. In fact, before antibiotics became available, doctors sometimes *induced* fever to treat a variety of infections, including syphilis. So don't be in a hurry to lower your fever, at least until you know what's causing it. If you do, you may be trading temporary comfort for long-term consequences. Masking your fever in the interest of a good thermometer number doesn't mean a thing if the *disease process* causing it continues unchecked.

Many people wrongly believe that fever is always due to a bacterial infection, and that any old antibiotic is all you need when your temperature goes up. That sometimes works, but if it does, you're just lucky! *Specific* bacteria respond to *specific* antibiotics. If you take "potluck" with one that was left over in your drug cabinet from the last time you were sick, and to which the particular bug infecting you this time is resistant, you will have lost valuable days pounding your target with the wrong ammunition. In the meantime, the underlying infection spreads unchecked. The wrong antibiotic not only leaves you untreated, it can also make other bacteria that may infect you sometime in the future resistant to it. So, never take an antibiotic on the *assumption* that it will always eradicate *any* infection. Remember, too, that infections (and fever) can also be due to *viruses*, and viruses do not respond to antibiotics. Neither will the fever they produce. Finally, your fever may not even be the result of an infection—either bacterial or viral. Temperature elevation accompanies a wide variety of noninfectious disorders—heart attacks, strokes, use of drugs, cancer, an overactive thyroid gland, autoimmune diseases, or damage or injury to body tissues (burns are a good example).

How to Treat the Fever Once You've Decided to Do So

The above notwithstanding, fever that's *too* high is not only uncomfortable, but potentially harmful. At some point it becomes important to lower it even before a final diagnosis has been made. Readings above 103 degrees in children can cause convulsions—always frightening and sometimes dangerous. Among the elderly, or those at any age who have been weakened by chronic disease, a persistent fever revs the biological motor to a level the body may not be able to tolerate. For example, in someone with angina pectoris due to narrowed coronary arteries, the increased heart rate caused by a fever means more work for the heart and this, in turn, may worsen the angina. Or if your cardiac muscle is weak, it will pump less well when it's beating too fast.

I recently asked several of my patients what they considered to be the best way to lower fever. Most of them chose aspirin; a few voted for acetaminophen (Tylenol, Panadol), and one fat gentleman said he "feeds" his. The correct answer is acetaminophen, *not aspirin*.

Acetaminophen acts directly on the temperature regulatory center in the brain to lower fever. Although it doesn't agree with some people, it is a safe and effective drug for most. Here are some specific reasons I prefer acetaminophen to aspirin when temperature needs to be lowered:

- It has few side effects other than drowsiness.
- It does not irritate the stomach—an important consideration to anyone with ulcers and to older persons in whom only a few aspirin tablets can cause bleeding from the upper intestinal tract.
- It does not "thin" the blood, so it's safer than aspirin if you're on anticoagulant therapy or have any bleeding or clotting problem.
- If you're asthmatic, acetaminophen won't throw your airways into spasm like aspirin can.
- It is not associated with Reye's syndrome, a potentially fatal neurological disorder that occurs in children who have the flu or chicken pox. (Aspirin is, so *never give aspirin to a child with a viral infection*.)
- The risk of death or permanent damage to major organs after an overdose of acetaminophen (deliberate or accidental) is, relatively speaking, not great.

- It will not interfere with any antigout medication you may be taking; aspirin can.

Acetaminophen is also superior, for the control of *fever,* to the nonsteroidal inflammatory agents (NSAIDs)—ibuprofen (Advil, Nuprin, Medeprin, Motrin), diflunisal (Dolobid), piroxicam (Feldene), naproxen (Naprosyn), indomethacin (Indocin), diclofenac (Voltaren), and others. The NSAIDs share many of aspirin's drawbacks and characteristics. Although they are more effective than aspirin for *pain* relief, and less irritating to the gut, they are not quite as good as acetaminophen for lowering fever.

A word of caution: When the temperature is very high, like 105 degrees or more (as occurs in heat stroke), you'll need more than just a drug to lower it. Fever of that magnitude requires immediate hospitalization and *rapid* cooling in order to prevent convulsions, coma, permanent brain damage, or even death.

How to Use Acetaminophen Properly

- Acetaminophen works within 20 minutes, and its effect lasts about 4 hours. Most adults prefer the 325 mg tablet, but other forms are available—drops, suppositories, an elixir, chewable tablets, syrup, or capsules. It also comes in an "extra-strength" 650 mg tablet and a weaker pediatric formulation of 80 mg. To lower fever, you'll need 650 mg—that is, 2 of the adult tablets or 1 extra-strength pill every 4 hours. *Do not take more than 12 tablets, or approximately 4,000 mg, in any 24-hour period.* (For children under the age of twelve, half that amount, 200 mg or 6 tablets a day, is the limit.) If this maximum dose doesn't lower the fever sufficiently, sponge with lukewarm water.
- Always look for the *expiration date* on the label whenever you buy a bottle of acetaminophen. Although this drug has a long shelf life, it doesn't retain its potency forever. I suggest you replace your home supply every 2 years regardless of how many you have left. Keep the bottle tightly shut and away from light.
- Don't take acetaminophen (or any drug, for that matter) any longer than you must. Chronic excessive use of acetaminophen can aggravate a kidney or liver problem. The more you take, the greater the risk. Also, if you're a heavy drinker, you're better off using aspirin to lower fever

because the combination of acetaminophen and alcohol can damage a liver already insulted by booze.

• *If you've accidentally taken more acetaminophen than you meant to, call your doctor right away.* In the meantime, have a swig of syrup of ipecac (see page 221) or get yourself to vomit some other way in order to rid your stomach of the drug. Then make for the nearest hospital emergency room as fast as you can so that your stomach can be pumped to prevent additional absorption of the drug.

In summary, when you want to lower your temperature, it's usually best to start with acetaminophen. How refreshing to call a doctor in the middle of the night because you have a fever, and be told, "Take two *Tylenol* and call me in the morning"!

FLU: Because You Refused the Vaccine!

Hundreds of different viruses can cause respiratory illnesses in man, ranging from a simple cold to viral pneumonia. One of these infections—influenza—can be prevented by a vaccine, yet relatively few people receive it. They can't be bothered, they can't afford it, or they're afraid of it! As a result, many thousands die unnecessarily every year from the flu. Being too "busy" to be vaccinated is a matter of personal priorities; being unable to afford it is a national disgrace; being afraid of vaccination is foolish—the result of a myth, and myths die hard. *There is absolutely no way the flu vaccine can give you the flu,* because it is made from an inactivated virus.

Sometimes patients reject my recommendations to be vaccinated because they are convinced that their last shot actually gave them the flu. In reality, they were already incubating the virus when they were being vaccinated, and were not yet protected. It takes about 2 to 3 weeks for the vaccine to confer immunity. So these persons became sick not from the vaccine but because it was given too late.

The Centers for Disease Control each year publishes its recommendations as to who should be vaccinated against influenza. Those at "high risk" for its complications include anyone over age sixty-five, and individuals with chronic disease who, should they come down with the flu, might become seriously ill. But I recommend the vaccine to *all* my patients regardless of age, sex, or the state of their health—unless, of course, they are allergic to chicken or eggs (in which the vaccine is prepared). The new, purified preparations rarely cause any significant side effects.

If you're healthy, were not vaccinated, and come down with the flu, it will usually last about 7 days and clear up on its own without any treatment. It's in the very young, the infirm, the elderly, and those suffering from some other chronic disease that this infection presents a threat and requires intervention. Therapy for flu is twofold. We first attack the *virus itself,* and at the same time, relieve the symptoms it causes.

The classic symptoms of the flu are fever (usually 102 degrees or more), headache, and discomfort in your joints, all followed by a cough. As soon as you suspect you have the flu (and preferably within 48 hours), take amantadine (Symmetrel), 100 mg a day for 10 days. It requires a prescription. Amantadine attacks the type A influenza virus, which is *always* present in the environment regardless of whether or not there is an epidemic. Type B flu, on the other hand, against which amantadine is not effective, appears in cycles of 5 or 6 years. Amantadine given in time will lower the fever, make you feel better, and markedly shorten the course of the illness. If you have any kidney trouble, discuss it with your doctor before taking this drug.

Amantadine not only effectively treats the flu, it also *prevents* it. So, if you haven't been vaccinated and you wake up one morning to find yourself in the midst of an epidemic, either in your neighborhood or at home (the kids, your spouse, your mother-in-law have all come down with it), don't just curse your luck. Get your vaccine shot right away, and at the same time start oral amantadine and continue it for the 2 to 3 weeks it takes for the vaccine to confer immunity. You may then stop the amantadine.

If you already have the *symptoms* of a full-blown flu, here's how to treat them. Start amantadine as discussed above. In addition, for the fever, aches, and pains, take acetaminophen (Tylenol). *Don't ever give aspirin to children with any viral illness* because of its association with Reye's syndrome, a serious neurological disorder. For cough control, use one of the preparations described on page 82 together with a decongestant recommended on page 81 There's no proof that antibiotics are of any benefit in influenza, except when there is some complication like pneumonia, and no doctor who is a pure scientist will ever prescribe an antibiotic for you. That's because it doesn't destroy the influenza virus. But those of us who have been treating flu patients for years will usually have you take an antibiotic anyway, especially if you're elderly and chronically sick. I generally prescribe ampicillin, 250 mg, 4 times a day.

In summary, influenza is not difficult to diagnose. It causes sudden high fever, headache, cough, and generalized aches and pains. Amantadine, an antiviral agent, will usually shorten the course and severity of this infection if taken in time—i.e., within 48 hours after the onset of symptoms. This drug is effective not only in the treatment of the disease, but also in its *prevention.* If you've not been vaccinated, do so as soon as you know you've been exposed or if there is an epidemic in your community, and begin amantadine at the same time. The vaccine confers immunity in 2 to 3 weeks, after which it's safe to stop the amantadine. I recommend vaccination annually for everyone who is not allergic to chicken or egg protein.

FROSTBITE: Putting Some Warmth into the Chill

Early every May, I make a special point of going to Yankee Stadium in New York to watch the home baseball team play one from Canada. Before the action starts, the band strikes up two national anthems—our own, and that of our neighbor to the north. Theirs begins with "Oh Canada, our home, our native land." The moment I hear those lugubrious strains (the Canadian national anthem is not the rousing, rhythmic, stirring kind of music to which you want to march), I relive scenes from my childhood in Canada where I was born—and I shiver! Particularly vivid are my memories of ice skating with my father in the dead of winter in Montreal. Thermometer numbers didn't mean much to me when I was five years old—but 20-degrees-below-zero weather often left me temporarily frozen during those outings. My ears, tip of the nose, toes, cheeks, and fingers would first redden to a healthy glow (that's *frostnip,* when you simply feel *very cold*), and then become waxy, white, and numb (that's *frostbite*).

My dad was not a doctor, but he knew exactly what to do when I was overexposed to the cold. First, he took me indoors as quickly as possible. He knew better than to let me smoke (possibly because I was only five years old), and never did so himself (perhaps because he also knew that cigarettes constrict blood vessels and so further decrease the amount of blood available to the frozen areas). Nor did either of us have any alcohol to "warm up," presumably because he was aware that alcohol shunts blood away from frozen tissues and aggravates the situation. Instead, he had me drink lots of hot *nonalcoholic* beverages. He made only one mistake in his cold-treatment regimen, something that puzzled me then and alarms me now even in retrospect. He actually rubbed my hands, ears, and nose with snow! My brother, with whom I reminisced the other day and who still lives in Montreal, rubs *his* frozen grandchildren the same way! It's madness! Don't *ever* do that no matter where you live. The treatment for any frost injury is heat, not more cold!

If you plan to be out in the cold, here's what to do. Try to anticipate and avoid frostbite by dressing adequately for the cold weather. Wear clothing in several layers, use mittens rather than gloves (so that the fingers keep each other warm), wear woolen socks, insulate your boots, skates, or ski boots and make sure they're not too tight, and cover your head (30 percent of heat lost from the body is from the head, especially if it's bald). And don't just stand around; keep moving about and drink hot liquids at least twice an hour.

The best *treatment* for frostbite, once you're out of the cold, is to immerse whatever part of you is frozen in warm (*not* hot) water. Remember, however, that these tissues are temporarily anesthetized, and don't accurately perceive hot and cold. So have someone else run the tap for you and decide on the proper temperature. Here's another important point: *Never thaw the affected limb before you arrive indoors.* This causes ice crystals in the frozen portion of the body to melt. When re-exposed to cold weather, the water they now contain refreezes, and this time can produce even more serious damage.

Freezing leads to two major vascular problems. First, blood circulating through the frozen area thickens and clots more easily. If frostbite is not treated in time, gangrene occasionally leading to the loss of a limb may result. Also, tiny ice crystals form around the blood vessels leaving them rigid and further injuring the affected area.

My earlier years as a doctor were spent in Montreal, the very city where I used to freeze time and again as a child. While serving my internship there, I was sent out in an ambulance almost daily during the winter to bring frozen skiers from the slopes of Mount Royal back to the hospital. (That's where I got the nickname Isadore Frozenfeld!) I learned then that *immediate* rewarming is the best way to avoid cold gangrene and all the other complications of exposure to cold sufficiently prolonged to produce tissue damage—like death due to electrical disturbances within the heart. Remember, frozen patients require the same meticulous care and monitoring we give to burn victims.

FUNGUS AND OTHER CONDITIONS OF THE MOUTH AND TONGUE: When You Look in the Mirror and Say "Aah"

If you're in the habit of examining your tongue every morning, here's what to do about some of the "abnormalities" you may discover. For example, is it covered with dark brown or black "hair," so that it looks more like a scalp than a tongue? If so, you have what doctors call "*black hairy tongue,*" probably due to antibiotics, chronic irritation of poorly fitting dentures, or excessive cigarette smoking. The best way to deal with it is, yes, brush the tongue twice a day, but with a toothbrush and not a hairbrush! Also rinse your mouth with hydrogen peroxide and avoid such irritants as alcohol and tobacco. If there are also colored wisps of hair "growing" out of your tongue, don't bother going to the barber. Just snip them off yourself with a pair of scissors! The bad taste associated with the hairy tongue is due to fermenting food particles trapped in its crevices.

Thrush (candida) is a fungal infection of the mouth and tongue that often follows long-term steroid or antibiotic therapy or radiation to the head and neck area for treatment of cancer. Candida appears as white patches in the mouth that look like cottage cheese and bleed when you try to wipe them away. The best way to get rid of thrush is first to stop the antibiotic or steroid (if it is safe to do so) and rinse your mouth with nystatin (Mycostatin) solution 3 to 4 times a day. If you prefer, you can swallow the tablet instead, 25 mg, 3 times a day for 7 to 10 days. If the fungus has extended down into the food pipe (esophagus), Mycostatin won't eradicate it. In such cases, I have found the antifungal agent ketoconazole (Nizoral) most effective. But remember when using this particular agent to have your liver checked every 2 to 4 weeks with a blood test.

Is your mouth chronically *dry* and uncomfortable? If the cause is not a local problem, like infection of the gums, teeth, dentures, or the salivary glands, it may be the result of some medication—a diuretic, an antihistamine, a blood-pressure-lowering agent, or an antispasmodic (to reduce hyperactivity of the bowel). When such an agent is identified, the obvious best treatment is to stop it or find a substitute. Other conditions that can give you a dry mouth include radiation therapy to the head and neck and certain diseases like rheumatoid arthritis, lupus, and Sjögren's syndrome.

Whatever the cause, tell your dentist about your dry mouth because lack of saliva predisposes you to dental cavities. Drinking more water does help. Also try using Xero-lube, an "artificial saliva," for relief; it's available without a prescription.

FUNGUS ON THE FEET: Are Yours Athletically Inclined?

When anything fungal hits the skin, we call it tinea. On the body as a whole, it is called tinea corporis, commonly known as ringworm; in the groin or thigh (a frequent location because of the moisture in the area), it is tinea cruris; on the feet you may call it athlete's foot but doctors refer to it as tinea pedis; on the head it's tinea capitis. Regardless of where you got your tinea, the treatment is usually the same—a topical antifungal cream or ointment. You won't often need an oral preparation.

In the old days, pharmacists were kept busy making up Whitfield's ointment, which most doctors prescribed for treating almost any skin fungus. Also, many of us grew up with the familiar orange tube of Desenex and the red and white container of Tinactin for our athlete's foot. These products are still available and effective, but the best treatments now are imidazole compounds that go under various trade names—clotrimazole (Lotrimin, Mycelex), miconazole (Monistat-derm), ciclopirox (Loprox), and others. Rub any one of these preparations into the affected area twice a day for 2 to 4 weeks. In addition, I still recommend dusting or spraying the affected areas with antifungal powders such as Tinactin or Micatin, morning and night, after showering.

If local treatment does not prevent the fungus from spreading, an oral preparation will. I recommend griseofulvin (Fulvicin), 500 mg, once or twice a day until relief is obtained. Ketoconazole (Nizoral) taken by mouth is also effective, but because of its potential liver toxicity, I do not prescribe it as a first-line drug for such a minor affliction.

It's interesting that although some members of my family have from time to time had athlete's foot and we walk the same floors, I have never acquired it. Some immunologists suspect that there is a component of personal or genetic susceptibility involved.

GALLSTONES: Suddenly Lots of Options

It wasn't so very long ago that anyone with gallstones was advised to have the gallbladder surgically removed as soon as possible regardless of whether or not the stones were causing any symptoms. The rationale for this aggressive approach was that the mere presence of stones meant the gallbladder was irreversibly diseased, and that "sooner or later" it would need to come out—so why not now? After all, most people are a better surgical risk today than they are apt to be when they're older.

That counsel would make eminently good sense were it not for the following facts: Fewer than 30 percent of persons with gallstones ever develop symptoms, and one attack does not necessarily mean that there will be another—provided you follow a low-fat diet. So, my advice to anybody whose gallstones were detected by chance—as, for example, in the course of an X ray, ultrasound, or CT scan of the abdomen taken for some other reason—and who is suffering no symptoms, is to leave them alone. However, if you are experiencing repeated acute attacks, the problem must be dealt with. Happily, there are several excellent options, most of which were not even on the drawing board just a few years ago.

In order to decide how best to deal with your gallstones, it helps to understand what the gallbladder is, where it's located, its function, why gallstones form, how they can possibly harm you, and under what circumstances.

One of the many functions of the liver is to make bile, which is necessary to digest fat in the diet. Bile leaves the liver via a duct (the common bile duct) that empties into the small intestine, where it acts on the fat you've eaten and digests it. Although the liver is continuously making bile, you're not always eating fat. So there are times when the bile entering the gut from the liver is wasted. In order to avoid that, nature has provided the common bile duct with a branch called the cystic duct. This leads into the gallbladder, situated very near the liver, and which is nothing more than a storage sac for bile. When there is no fat in the small intestine, and thus no need for the bile the liver is producing, it is diverted from the common duct to the cystic duct and into the gallbladder. Later, after you've indulged in some butter, milk, cream, cheese, marbleized steak, eggs, and other foods calculated to give the American Heart Association a fit, the intestine sends a signal calling for more bile. The gallbladder responds by squirting out what it has been storing, through the cystic duct and into the fat-filled small intestine.

Normally, the bile in the gallbladder is in totally liquid form. Sometimes, however, its constituents, the most important of which is cholesterol, come out of solution. In other words, they form little crystals. These gradually increase in size to become stones. It's not entirely clear why that happens. A low-grade infection of the gallbladder is one factor; the use of estrogens is another; genetic influences play a role, too. Regardless of the mechanism responsible for their formation, gallstones are not a good thing to have. Most of the problems they cause result from their leaving the gallbladder and entering the ducts. If they happen to be stuck in one of them, you will experience severe pain in the right upper abdomen that comes in waves as the duct contracts to squeeze the stone down and out. If the stone gets into the common duct, and obstructs it, bile can't get out of the liver, backs up, eventually ending up in the bloodstream and leaving you jaundiced. (Bile contains the yellow pigment called bilirubin that causes jaundice.) The smaller the gallstones, the more likely you are to have trouble because larger ones are too big to slip into the bile ducts. So stones just sitting in the gallbladder don't usually give symptoms. All this explains why it's important for anyone with gallstones to follow a low-fat diet, because if you don't, the fat you eat causes the small intestine to send a message to the gallbladder calling for more bile. When it responds by contracting to provide the needed additional amount, a stone or two may slip out with the bile and voilà, you have an attack.

Surgery remains the "gold standard," the one definitive way to deal with gallstones. If you are young and healthy, and they're troubling you, that's the route to go. However, once the gallbladder and the stones it contains are removed, there is no guarantee that your symptoms will not reappear in the future. Years later, you may form new stones, this time not in the gallbladder (which is now gone) but in the ducts that carry the bile from the liver to the intestine. Actually, in one of every hundred gallbladder operations, stones are *already present* in those ducts, but are undetected. That's what happened to the late Shah of Iran, who had his gallbladder "successfully" removed but shortly thereafter suffered a recurrence of his symptoms because several small stones had been left behind in the bile ducts!

Conventional gallbladder surgery is not a dangerous procedure in otherwise fit individuals, but the risk increases as you get older, especially when the operation is performed as an emergency rather than at a time and place of your choosing. This is particularly true if you also have other problems such as angina, heart failure, chronic lung disease, kidney trouble, liver malfunction, and so on. The overall death rate from such surgery is less than 2 percent, but there are potential, usually nonfatal, complications such as infection and blood clots. What's more, the operation hurts, it's costly, and you lose time from work (anywhere from 4 to 6 weeks, depending on how eager you are to return to the job). So, surgery is great if you need it because of symptoms, but otherwise avoid it.

Now here's some good news. In 1988, a surgical team in Nashville refined a technique developed earlier in France permitting the removal of the gallbladder by laparoscopic surgery. In my opinion, this operation has revolutionized gallbladder surgery. It is the *only* way I would have my own gallbladder taken out now if I needed to. In this technique (done under general anesthesia), four very small incisions (approximately one quarter inch long) are made in the abdominal wall. An instrument (a laparoscope) containing a miniature high-resolution TV camera is introduced into one of them, which permits the surgeon to see the gallbladder. A laser probe then dissects away the gallbladder, which is removed through one of the other small incisions. The operation takes about 1½ hours and is virtually bloodless. You can eat that evening, go home the next day, rest for a week, and then resume your normal activities—cured. Any surgeon who tells you to stay with the conventional procedure and "not take chances" with this new one probably doesn't know how to do it. Find someone who does, but stay away from a surgeon who has just come back from a weekend course learning about it; unless it's done the right way, complications from the "simple" procedure can be serious.

If you have decided against surgery—either because you don't like operations or you think you're too old and not in such great shape—what are your alternatives? When I wrote my book *Second Opinion* in 1982, there was only one—a medication to dissolve the stones. The preparation available then wasn't very effective, it had troublesome side effects, and the stones usually recurred after you stopped taking the drug. But there is a new agent, ursodiol (Actigall), which is far superior and which you should consider if you are not a surgical candidate. However, it will only dissolve cholesterol stones (surgery, of course, eliminates every kind), not those formed from pigments or calcium. Fortunately, 80 percent of gallstones in this country are composed of cholesterol, so this drug is suitable for most people. Your doctor can determine the makeup of the stone from its appearance on the X ray film and/or sonogram. Actigall will completely dissolve cholesterol stones in a third of cases if taken faithfully for 1 to 2 years. The best results are obtained when there are many small stones, rather than just a couple of large ones. It's also very good when "sludge," rather than stones, is the problem. The dosage of 2 or 3 tablets a day (with meals) now costs about $1,000 a year, plus the charge for the ultrasound examination of the gallbladder that is required twice a year to make sure all the stones are gone. If they are, continue the medicine for another 3 months, then stop it. In any case, Actigall should not be taken for more than 2 years. Although side effects from Actigall are rarely troublesome enough to warrant stopping it, you should, nevertheless, have routine liver tests while on this drug because it can affect the liver. Also, while taking Actigall, you should avoid cholestyramine (Questran, for lowering cholesterol) or any antacid containing aluminum, both of which interfere with the absorption of the Actigall.

If the Actigall dissolves all your stones, there is still a fifty/fifty chance that they will form again within 5 years. Should that happen, your choices now are to throw in the towel and have the gallbladder removed, or try a second course of the drug. The latter is what I would do if it was successful the first time around.

There now are two other approaches to the management of gallstones, both of which are still experimental. Neither, in my opinion, has any advantage over the laparoscopic surgery. In the first technique, pioneered by Dr. John Thistle at the Mayo Clinic, the abdominal wall is pierced by a needle that is then introduced directly into the stone-filled gallbladder. An ether solution injected through the needle dissolves most of the stones within 12 hours. This treatment requires hospitalization for 2 or 3 days. The success rate is said to be greater than 90 percent. Several of my own patients with gallstones in whom surgery was too risky, or who failed treatment with Actigall, have responded to this ether therapy although tiny stone fragments sometimes remained. Ether, like Actigall, dissolves only cholesterol stones, but here the size of the stone is not a factor.

Shock wave lithotripsy (the word *lithotripsy* means "stone crushing") is yet another treatment option for gallstones. This technique is very similar to that used for shattering kidney stones. An underwater electrode discharges a high-voltage spark which causes the sudden evaporation and expansion of water in a water pillow on which the patient is lying. The resulting shock wave is directed under ultrasound control to the stones in the gallbladder. Fragments of the shattered stones are passed out through the bile ducts during the next few days.

I've personally observed gallstone lithotripsy in the Munich Hospital where the technique was developed, and although it does crush the stones, there are several limitations to its use. The gallbladder itself, despite the stones it contains, must be functioning normally, and that's not always the case; lithotripsy will only shatter those stones composed of cholesterol and containing no calcium; and the treatment should not, in my opinion, be administered during an acute gallbladder attack or to anyone who has a cardiac pacemaker. In order to prevent recurrence, patients undergoing gallbladder lithotripsy must take Actigall a week before their "shattering" experience, and for about 3 or 4 months thereafter.

Lithotripsy does not require full anesthesia, but sedatives are given because the procedure is uncomfortable. In one of the cases I witnessed, the patient received 1,400 shocks in less than an hour to fragment her one large stone. Believe me, she really appreciated those sedatives. Several centers in the United States are now evaluating gallstone lithotripsy, and the final verdict on its overall usefulness should soon be in. It is not my number-one recommendation at this time.

In summary, it is not usually necessary to remove gallstones just because they are there. The likelihood of your developing a problem with such silent stones in the future is not great, especially if you continue to follow a low-fat diet. However, if you're having recurrent attacks of pain and fever, and especially if you're otherwise in good health, then the best long-term results are obtained with surgery. If at all possible, have that done with the new laparoscopic method rather than the conventional procedure. If you are a poor operative risk and the need for surgery is not urgent, then dissolving the stones (provided they are composed of cholesterol) with Actigall is the course I recommend. Expect to take this drug for as long as 2 years. If the situation does not permit treatment for this length of time and surgery is chancy, contact the Mayo Clinic and ask whether you are a candidate for their ether injections. If you're not, then the still-experimental lithotripsy is a final option.

GAS IN THE GUT: More Than Beans

A certain amount of gas is normally present in the gut, resulting in an occasional, harmless, surreptitious southerly release when it's low in the bowel, or a discreet burp from higher up in the stomach. Here's a research figure that may surprise you. (Give a scientist a laboratory and a grant, and he'll surely come up with *something* to investigate.) Healthy individuals who eat a "normal" diet pass gas from the rectum ten to eighteen times a day! Now, if this event is silent and odorless, or if you're alone at the time—at home, in your car, on the street—there's no problem. But if you're noisy about it, or have been eating certain foods that are converted into the very aromatic methane gas in your bowel, you and those around you will be embarrassed and uncomfortable, respectively. Although belching is generally viewed as less offensive, it is not exactly one of the social graces either.

If normal individuals "break wind" ten to eighteen times a day, imagine the statistics for those who actually complain of *excessive* gas! I just happen to have those numbers for you, too. The answer is 140 times a day or about six times an hour! (Now these people really have it rough, passing wind once every 10 minutes!)

In most cases, excessive air in the intestine gets there either because you've swallowed it or because bacteria in the bowel are producing it in larger than normal amounts. Everyone swallows *some* air when eating, but nervous individuals gulp it down in huge quantities. But you may also swallow too much air even if you're calm and relaxed. For example, when dentures don't fit properly, lots of saliva is produced which is then swallowed along with air; nausea also induces more frequent swallowing, again causing more air than normal to enter your stomach; drinking liquids through a straw, chewing gum, sucking candy, smoking, quenching your thirst with carbonated beverages, or taking baking soda for indigestion will all give you an abnormally high gas content in your intestinal tract.

What happens to all this gas? If you sit up, you'll belch it out. If you lie down, the gas moves into the bowel and is expelled as flatus. The word *fart* has never been accepted as sufficiently dignified to be printed in the American medical literature, although it is standard usage by our British colleagues. However, one of my patients was not surprised when he came across *fart* in a report he received from me. As part of a complete evaluation, I had drawn a diagnostic battery of blood tests. In reviewing the results, I noted that the sugar was a little high. I couldn't remember whether this man had eaten breakfast before coming to the office that morning, and so I added a small note in my own handwriting at the bottom of the report. "P.S. Were you fasting when this blood was drawn?" Apparently he did not find my script legible because he mistook my *s* for an *r*. He called to ask me how on earth I knew—and incidentally, what difference did it make?

When excessive gas in the gut is not due to air swallowing, the most likely cause is fermentation by normal bacteria in the bowel processing foods like cabbage, cauliflower, baked beans ("the musical fruit, the more you eat, the more you toot"), peas, sprouts, peaches, apples, melons, and onions. But, one person in three harbors the kind of bacteria that ferment virtually *every* food and generate large amounts of foul-smelling gas with the odor of rotten eggs. This characteristic runs in families and becomes apparent in infancy to parents with an intact sense of smell. Another cause of excessive gas in the gut is suddenly adding large amounts of insoluble fiber such as bran to your diet instead of increasing its intake gradually. The shocked bowel reacts by forming too much gas in the intestine.

In addition to all the "normal variants" that cause bloating, there are some intestinal disorders that do so, too. The most important are an irritable or spastic bowel (see IRRITABLE BOWEL SYNDROME), gallbladder disease (probably because the fat you eat is poorly digested), and lactose intolerance, in which there is a deficiency of the lactase enzyme required to digest lactose (the sugar that is present in milk and milk products), and the undigested sugars are then fermented by bacteria and produce gas.

Is there a way to reduce the amount of gas in your gut? Unfortunately for you and those closest to you, there is no medical treatment that does so predictably. There are, however, some steps you can take to minimize the problem.

- Avoid those foods that you know from experience cause *you* bloating.
- If you're an air swallower, eat more slowly. For those nervous individuals who gulp air all day, holding a pencil between the teeth reduces the amount swallowed.
- Don't drink liquids through straws.
- Stop smoking.
- Avoid carbonated beverages.
- Don't suck on candies or chew gum.
- Reduce the amount of nonabsorbable fiber such as bran in your diet.

Most medications touted as antiflatulents don't really work. You will see and hear advertisements claiming that simethicone, especially when combined with antacids, reduces gas formation by breaking up small accumulations of gas or causing them to coalesce into larger bubbles than you can then burp out. Among the products containing this substance are Mylicon, Mylicon-80 (a more potent version), or Silane. Try them, but don't expect complete relief. Pancreatic enzymes are another frequently recommended antiflatulent, although they help only when you have been shown to be deficient in them—and that's quite uncommon. *Activated charcoal* is useful in many cases of poisoning but has no effect on gas. When taken regularly or frequently, charcoal can interfere with the absorption of medications and nutrients that you really need. In my experience, *antacids* don't provide relief either.

So what's the bottom line? If you have a chronically distended belly accompanied by troublesome belching and/or farting, it's a good idea to tell your doctor about your problem on the outside chance that you're one of the few with a treatable disorder. (One of my own patients actually insisted on demonstrating his complaint in a private "concert," while another brought me a tape recording!) The most likely culprit, statistically speaking, is lactose intolerance, in which event the milk you drink should be pretreated with Lactaid, and you must either eliminate lactose from your diet or take lactase pills to replace the missing enzyme.

GERMAN MEASLES: Mild Symptoms with Lethal Consequences

German measles (rubella) is a viral disease so mild there is no treatment required or available. But it can have serious consequences when the wrong person contracts it at the wrong time.

If your child complains of mild headache, sore throat, and a "cold," all of which are followed in 2 to 3 weeks by a rash on the face that later spreads to the arms and the rest of the body, chances are it's German measles. Swollen glands almost always appear on the back of the neck as well. The rash clears in a few days—and that's all there is to it. If the child complains of aches and pains, acetaminophen (Tylenol) will be enough to control them. *Don't ever give aspirin to kids with German measles, or any viral infection,* because of its connection with Reyes syndrome, a serious neurological disorder.

Remember, you *want* your child to be immune to German measles. The best way is vaccination, failing which you should arrange deliberate exposure to someone who has it. That's especially important for your daughter, so that if and when she becomes pregnant later in life, she will have acquired immunity. A woman who develops German measles in the first 4 weeks of pregnancy runs a 60 percent risk of giving birth to a baby with serious congenital defects. Infection at that time is grounds for abortion. If, however, you have strong religious or moral qualms about terminating the pregnancy, you should at least obtain immune globulin by injection as quickly as possible. This may lessen the likelihood of congenital malformations in the fetus, but you can't count on preventing them.

If you become pregnant, have never received the German measles vaccine (which every child should obtain at the age of fifteen months together with measles and mumps vaccine), and can't remember whether or not you were ever infected, there is a blood test that can detect the presence of German measles antibodies. If you are planning to have a baby but are *definitely* not yet pregnant and have none of the antibodies, get yourself vaccinated *before trying to conceive* and then don't get pregnant for 3 months following vaccination. However, since this is a live vaccine, do not take it if your immune system (your resistance) is impaired, as for example in cancer, or you are receiving steroids for any reason. Under these circumstances, the live vaccine can actually infect you. Also, if you were given a blood transfusion or a gamma globulin shot in the preceding 3 months, hold off receiving the vaccine. Somebody else's antibodies present in the blood or blood products you were given may inactivate the virus in the vaccine that is supposed to stimulate *your* immune system.

GONORRHEA: It Keeps Bad Company

I remember the days when patients who'd been out on a spree would stop by the emergency room for a quick shot of penicillin just in case they'd picked up gonorrhea. "A dose" or "the clap," as it was affectionately called, was not considered a major problem then because it was so easily treated. If one were unlucky enough to have contracted syphilis during the same encounter (gonorrhea and syphilis were the sexually transmitted infections one heard most about years ago), then the penicillin shot would take care of that, too. The specter of AIDS was not yet present; homosexuality was still in the closet; nobody had even heard of chlamydia; herpes was not on anyone's mind; and penicillin-resistant gonorrhea had not yet emerged.

How times have changed! If you are promiscuous—single, married, heterosexual, bisexual, or homosexual—you had better take gonorrhea seriously. Although chances are that one shot of penicillin will still probably cure your infection, you can no longer count on it. More and more strains of gonococcus, the organism that causes this disease, have become resistant to this antibiotic. Gonorrhea that has not been eradicated can lead to several major complications ranging from pelvic inflammatory disease in women and chronic prostate infection in men, to severe arthritis due to dissemination of the organism throughout the body in both sexes.

One gonorrhea case in three in many major cities in this country, and indeed throughout the world, is now penicillin-resistant. So, instead of penicillin, I give my patients an injection of ceftriaxone (Rocephin), 250 mg, into the muscle. Unfortunately, this antibiotic costs ten to fifteen times more than the equivalent 1,200,000 units of penicillin that will cure nonresistant organisms. In addition, I prescribe 7 days of tetracycline, 500 mg, 4 times a day, or doxycycline (Vibramycin), 100 mg, twice a day in order to eradicate chlamydia, which is the handmaiden to the gonorrhea bug in almost half the cases. (Homosexuals don't need tetracycline because chlamydia is spread between the sexes, and not from man to man.)

If you've acquired *any* sexually transmitted disease, make sure that your partner is also treated, regardless of who gave what to whom, and where the infection was contracted. Finally, men indulging in casual sex should *always wear a condom,* and women should insist that their lovers do.

GOUT: It's Usually a Pain in the—Toe!

The other day one of my patients phoned to inform me that he had consulted a "foot specialist" about a painful bunion, and that arrangements had been made for its early removal. The specialist told him that the bunion was the result of his having worn narrow shoes for so many years, and that surgery was now his only option. When I asked how long he'd had the pain, he told me that his foot had suddenly started "hurting like hell" a couple of days earlier. It was now so tender he couldn't even cover his feet with a bedsheet at night. To me, that description was the tipoff to gout. Bunions don't suddenly become so exquisitely painful that you can't even brush lightly against them. I asked my patient to stop by later in the day to let me have a look at his foot. I found the toe swollen, red, warm, and extremely tender. This was gout, not a bunion! My patient would have been in trouble had that acutely inflamed toe been operated on.

Gout is a very common disorder that primarily affects men. It has been described in literature, including the Bible, for millennia. Some very famous people have had it, like Ben Franklin (who didn't get it from flying kites), Alexander Hamilton (it wasn't from a Burr), Isaac Newton (the falling apple had nothing to do with it), Achilles (who did not have it in his heel), and Oedipus (who probably did not inherit it from his mother). It can affect almost any one of the smaller joints, but most frequently strikes the great toe. This special predilection is why gout is still sometimes referred to as podagra, from the Greek phrase "foot caught in a trap." (Larger joints, such as the back or hips, are rarely if ever attacked by gout.)

Gout is the result of too much uric acid circulating in the blood, either because of overproduction by the body (as in certain malignancies like leukemia) or not getting rid of it fast enough (as occurs when the kidneys are diseased). Uric acid will also be elevated by diuretics.

Whatever the reason, when there is more than a normal amount of uric acid in the blood, its crystals settle down in the joints and inflame them. If you want some idea of how painful that can be, imagine dumping a whole lot of these needle-like crystals inside an elbow, toe, finger, knee, or wrist and then trying to touch or move it!

There is no cure for gout. The painful attacks can be treated, and their recurrences prevented, but the underlying predisposition is permanent. The key to *preventing* attacks of gout lies in taking medication on an ongoing basis to lower the elevated uric acid. When this increase results from your body's inability to eliminate uric acid, I recommend sulfinpyrazone (Anturane)—200 to 400 mg daily in divided doses. This drug helps the kidney excrete it in the urine. Aspirin interferes with the beneficial action of Anturane, so avoid it when taking that medication. When the elevated uric acid is due to its exessive production, I recommend a single 300 mg tablet of allopurinol (Zyloprim) every day, to control its manufacture.

Both Zyloprim and Anturane are usually well tolerated, but occasionally cause a skin rash, and Zyloprim sometimes induces altered liver function. Have a routine blood test once or twice a year if you're taking it. If despite Anturane or Zyloprim the attacks of gout keep recurring, add 1 or 2 pills of colchicine (0.6 mg) to your preventive regimen every day. This agent has a specific action on the acutely inflamed joints in this disorder.

During an acute attack of gout itself, however, neither Anturane nor Zyloprim should be used. If your gout breaks through while on these preventive drugs, stop taking them and start one of the anti-inflammatory agents discussed below.

Supposing, despite your best efforts at prevention, you awaken one morning with an exquisitely painful toe and you need relief fast. Here's what to do. First, stop the preventive drugs (Anturane or Zyloprim). Then, start indomethacin (Indocin), 50 mg, 3 or 4 times a day after meals; it will stop the pain and reduce the inflammation within 24 hours, but must be continued for 3 days after all symptoms have subsided. If you are also on anticoagulants, check your blood frequently because Indocin increases the effect of the anticoagulant and can cause bleeding. About 2 weeks after the pain has gone, return to your preventive regimen of Zyloprim or Anturane, but in addition, it's a good idea to add 1 or 2 tablets of colchicine a day for 3 or 4 months to reduce the likelihood of a recurrence.

Before the availability of specific medications to treat and prevent gout, patients with this disorder were forbidden to have alcohol, caffeine, anchovies, meat, and animal organs like kidney, liver, brains, sweetbreads, and minced meats. Vegetables like asparagus and mushrooms, which raise the uric acid level, were also dietary no-no's. Failure to comply with these prohibitions often led to a very painful siege. But even in the most cooperative individuals who were extremely careful with their diets attacks were still induced, by acute stress—profound anxiety, fear of impending surgery—infection, diuretics—or for no apparent reason. Regardless of the mechanism that induces gout in any particular case, unless the specific medications now available to control the production and excretion of uric acid don't work, and that's highly unusual, I do not insist on rigid adherence to the dietary restrictions mentioned above. As far as social drinking is concerned, a couple of beers or glasses of wine or two cocktails in the evening are usually well tolerated. Here's a final piece of good news. According to recent reports, frequent orgasms in men lower the uric acid level in the blood and decrease the frequency of attacks!

HEART ATTACK: What *You* Know and Do Can Make the Difference Between Life and Death

More than half of the million people who suffer heart attacks every year in this country never even make it to the hospital. They die suddenly—at home, at work, or at play—often without warning, before any treatment can be started. The only way to deal with sudden cardiac death is to prevent it, and at the moment that's best done by eliminating the major risk factors associated with arteriosclerosis—high blood pressure (see page 132), cigarette smoking, and elevated blood cholesterol levels (see page 59).

For the 500,000 individuals who do survive a heart attack long enough to be helped, the best treatment requires several steps to be taken. The first and most important is to be prepared. *Every adult should be able to recognize the earliest symptoms of a heart attack,* and to distinguish them from indigestion, hiatus hernia, gas, nerves, and the other common disorders responsible for discomfort in the chest area. A heart attack usually causes a feeling of heaviness, tightness, or pressure—not necessarily pain—behind the breastbone in the *center* of the chest, not by the left nipple. The discomfort can also radiate to the teeth, the jaws, both shoulders, the elbows, or the back, and it is generally accompanied by a cold sweat and some shortness of breath.

Here's an example of how this kind of information can save your life. One evening, a fifty-year-old friend of mine developed a cold sweat and a "heavy pressure" behind the breastbone. Suspecting a heart attack, he immediately called 911 and was rushed by ambulance to a nearby hospital emergency room. By the time he arrived, his symptoms had abated and his electrocardiogram was found to be normal. The resident physician on duty, for whom at least ten other very sick people were waiting, reassured the patient that he was "fine," that his ECG was "perfect," and that this was just an attack of indigestion. He was free to leave. But this gentleman, in addition to recognizing the classic symptoms of a heart attack, also knew that a normal ECG does not exclude that possibility. So he refused to go home. After vacating the examination cubicle, which was needed for the next patient, he took a seat in the waiting area. A half hour later, while he was witness to several other dramas unfolding in the hectic emergency room, his chest pain recurred. This time the ECG did indicate an acute heart attack, and the chagrined resident immediately initiated emergency treatment. Later, the patient who had "cried wolf" was transferred to the coronary care unit, and was discharged from the hospital in good condition after 8 days. Had this man gone home when advised to do so by the harried doctor, he might well have died. This true account underlines the fact that if you suspect that you may be having a heart attack, get help fast, and don't take a quick no for an answer.

If you develop symptoms that you believe may indicate a heart attack, making the right decision and acting swiftly is critical. Do not waste time trying to locate your doctor if he's not in his office. Every minute is precious. He may be away and not have told anyone where he can be reached; the battery on his beeper may be dead; he may have gotten your message but be unable to respond immediately because he's caught in traffic and doesn't have a phone in his car; the doctor to whom he is "signed out" may not be conscientious about responding quickly to someone else's patient. The list of possibilities that can prevent you from making instant contact with your doctor are endless. So, here's *rule number one:* If you *suspect* you're having a heart attack, particularly if you're vulnerable—because you've had one before, or your family is riddled with heart disease, or because you have one or more of the risk factors associated with arteriosclerosis—call 911, or whatever the emergency number happens to be where you live, immediately for an ambulance, preferably one with trained paramedics. (Remember, you don't have to be sure you're really having an attack to make that call.)

Decisions You Should Make in Advance

• *Do you know what hospital to go to in the event of a cardiac emergency,* and the fastest way to get there? If you don't, then put this book down, think about it, and discuss it with your family—now. That's a key decision that should be made while you're well, not in a crisis situation. In choosing an emergency room, keep the following facts in mind. It's usually advisable to go to the hospital where your own doctor can participate in your care. But the emergency vehicle responding to your call will, in most cases, transport you to the *nearest* facility, and the paramedics will not be moved by your pleas to be taken to your doctor's hospital, especially if it's any distance away. So you should keep handy the telephone number of the ambulance company that services the hospital of your choice, and call them in an emergency, if the 911 squad is not likely to transport you where you prefer to be taken.

The most important consideration in selecting a hospital is the quality of cardiac care it can deliver. There have been many recent advances in the treatment of acute myocardial infarction (that's what doctors call a heart attack) of which you should take advantage. For example, when I started practice, mortality from acute coronaries at the New York Hospital Medical Center, my own hospital, was about 30 percent—and it is one of the best cardiac centers in the country. That number is currently only 5 or 6 percent. These figures are not the result of good luck, but reflect the impact of new drugs, modern coronary care units, better ambulance service manned by efficient paramedics, sophisticated diagnostic techniques available around the clock, 7 days a week, clot-dissolving drugs, and the possibility of performing early angioplasty or emergency cardiac surgery even during the heart attack itself. So whenever possible, go directly to a center where you can benefit from all these advances rather than to a smaller hospital from which you may have to be transferred at the most critical time of your life. If your own doctor is not on the staff of such a hospital, you should find a cardiologist who is—*before* you need him. If you're at risk, and what adult isn't, discuss the matter frankly with your family doctor and work out *in advance* a plan of action in the event of a cardiac emergency.

• *Choosing the right ambulance service is important, too.* Here's why. The modern ambulance is really a mobile coronary care unit. Remember that *you are at greatest risk in the first few minutes after the onset of symptoms,* during which time there are two treatment objectives. The first is to *correct any dangerous disturbance of heart rhythm that may be present.* Sudden death occurs when the heart goes electrically "haywire" in the very early stages of a myocardial infarction. Most paramedics are trained to recognize a life-threatening rhythm and treat it in the ambulance. In the past, that meant administering a drug called lidocaine intravenously. Although that's often effective, shocking the heart electrically with a *defibrillator* results in a more immediate and predictable response. Most ambulances now have two-way radio or telephone communication with the "base"—that is, the emergency room to which they're proceeding—so that the paramedics can consult with the doctor on duty before administering the shock. Some even have a combined electrocardiograph-defibrillator with a built-in computer, which identifies these life-threatening rhythms and *automatically* jolts the heart back to normal without the need to communicate with the base! There is yet a newer, more impressive piece of life-saving equipment, an ECG-defibrillator, which is hooked up to a cellular phone in the ambulance. The patient's electrocardiogram is transmitted by phone to the emergency room where the physician on duty makes the decision whether or not to shock. If necessary, he can push a button *in the hospital* that will activate the defibrillator in the ambulance via the telephone! Several hospitals have purchased these units from Medphone, a New Jersey-based company that has developed the technology. In any event, select in advance an ambulance service that has the capability of correcting a dangerous rhythm en route to the emergency room.

After dealing with the immediate threat of a chaotic heart rhythm, there is a second major decision to be made: *Whether or not to administer clot-dissolving medication.* Most heart attacks occur as the result of the complete closure of one or more of the coronary arteries. This cuts off the blood flow to the portion of the heart muscle they supply. The size of the artery involved, and how long it has been occluded, determine how much heart muscle will "die." That amount is the key to whether you will survive, and if you do, the quality of life you can expect later on. These arteries have usually been narrowing gradually for years, but at the final moment of the acute heart attack itself, they are completely shut down by a fresh clot. There are several drugs, called *thrombolytic agents*, that can quickly dissolve these clots and so reduce the length of time during which the heart muscle is deprived of oxygen. So, going to a hospital where this type of medication is available is critical. The earlier you are given a thrombolytic medication (t-PA, streptokinase, urokinase), the better off you are. For maximum benefit you should receive it within 5 or 6 hours at the very latest after the onset of symptoms. Ideally, a clot-dissolving agent should be administered in the ambulance—assuming, of course, that the diagnosis of a heart attack can be made there *with certainty.* Such a diagnosis is based on an interpretation of the symptoms, certain physical findings like blood pressure, and the presence of electrocardiographic abnormalities. If there is any question, you shouldn't have the drug since its use involves some risk. In most cases it's given in the emergency room, not in the ambulance. This is another reason to select a hospital in advance if possible. If the hospital to which you have been brought does not stock any of these drugs, too much time is lost (not to mention the risk involved in the additional transportation) sending you elsewhere for it.

What to Expect If You Have a Heart Attack

Let's assume, then, that you have correctly recognized the symptoms of a heart attack; you've discussed beforehand with your doctor where you will be taken in the event of a cardiac emergency, and you've planned all the arrangements for transportation. You finally reach the emergency room. If you've arrived on your own—that is, other than by ambulance—don't be bashful about telling the screening (triage) personnel that you have chest symptoms and think you're having a heart attack. If you do, testing and treatment will begin immediately. But unless you accurately, forcefully, and promptly describe your symptoms and indicate what you suspect, you may be kept waiting for hours and might just as well have stayed at home.

After one of the health care providers, usually a screening nurse, listens to your story, you will be examined, an electrocardiogram will be taken, and blood will be drawn for analysis. While the diagnosis of a heart attack is being confirmed, you will be given oxygen through two nasal prongs. A stand with a bottle of fluid (consisting of sugar and water) will be set up and a needle inserted into your vein. That's done in order to be able to inject any drug you may need later on quickly and directly into the bloodstream. If you're in pain, you will be given morphine. *Pain control is very important.* Don't grin and bear it; this is no time to be a hero. Severe pain can raise your pulse rate and affect the blood pressure. Anxiety doesn't help either, and there's nothing that makes one more anxious than severe pain. If you have a rhythm disturbance, you'll either be given lidocaine through the intravenous line to suppress the "extra beats" or, if necessary, you will receive an electrical jolt to correct the disorder.

Once the heart attack diagnosis is confirmed, you will likely be given a clot-dissolving medication. There are several from which to choose, but these days cost will determine the one you will receive. All these agents carry a small but definite risk of hemorrhage. The most expensive of the lot—by far—is t-PA. I used to think it was so much safer and more effective than the others that the price difference was worth it; however, recent studies have shown this not to be so. Chances are you'll be given whatever preparation the hospital happens to stock, and you will have no say in that decision. In addition to clot-dissolving medication, you'll also get, believe it or not, the lowly aspirin, which enhances its action.

At this point, you'll be transferred to the coronary care unit, where your subsequent treatment will depend on your condition. The most common complication of a heart attack is persistent pain, requiring repeated shots of morphine, an indication that the blocked artery is in spasm and that additional heart muscle is being damaged. An immediate assessment is now required as to whether or not you should have an emergency coronary angiogram. This is a good time to ask a cardiologist to consult with the interns and residents who, though skilled and dedicated as they are, lack the experience of the specialist. The right decision at this juncture can make the difference between life and death, and also determine the quality of your life if you survive.

If the major problem is not pain but recurrence of a dangerous rhythm disturbance, a brief burst of electrical energy to shock your heart may be required, or you may be given one of a wide array of drugs for that purpose.

The heart attack, especially if it's a bad one, may also be accompanied by a significant and continuing drop in blood pressure. That's probably the most serious complication there is, and requires either medication to raise the pressure or, in many cases, emergency bypass surgery to ensure survival.

To summarize, the three major complications that will determine your care and outlook after a heart attack are *persistent pain, cardiac arrhythmias,* and a *drop in blood pressure*. Regardless of whether or not you experience any of these complications early on, you should be admitted to the coronary care unit where your "vital" signs—heart rate and rhythm, blood pressure, respiration, temperature, and other bodily functions—can be monitored and treated immediately if necessary. In my opinion, no one with an acute heart attack should be luxuriating in a private room somewhere on a noncardiac floor of the hospital, even though such quarters are more comfortable than the usually more frenetic cardiac care unit.

If you're lucky, you will be one of the great majority of heart attack patients whose course is uncomplicated. Standard treatment in such cases consists of short-term blood thinning with heparin (an anticoagulant), intravenous nitroglycerin, beta-blockers, and/or calcium channel blockers (see discussion that begins on page 136). Which agent(s) to use will depend on a variety of findings at the time, and that decision will be made for you.

If all goes well, as it usually does, you will remain in the coronary care unit for about 48 hours and then be transferred to a "step-down" area where the monitoring is continued for a few more days. More and more institutions have established these intermediate facilities to provide additional protection to the patient in the event that there is any recurrence of life-threatening complications. In the step-down unit, your heart rhythm can be watched as you wear a wireless electrocardiograph that transmits your ECG to screens in the nursing area. A serious rhythm disturbance generates an alarm to which the staff can respond instantly. While on this floor, you will gradually increase your level of activity until it is safe to be moved to a regular hospital room.

Heart patients under my care who have had uncomplicated attacks usually remain in hospital no more than 7 or 8 days. Before discharge I often arrange for them to perform a very low level stress test to see whether or not they need angiography and angioplasty (ballooning open the artery) or bypass surgery (see page 20). The stress test also predicts potentially serious cardiac rhythm disorders that require medication in addition to those usually given upon discharge. The latter include 1 aspirin a day, a beta-blocker such as atenolol (Tenormin), whose ongoing use for a year or 2 has been shown to reduce the incidence of recurrent attacks, and blood pressure pills if your pressure is high. In addition, you will be given a low-cholesterol, low-saturated-fat diet—and possibly cholesterol-lowering drugs, too, if your blood fats are really out of whack—and an exercise program. The latter is important for two reasons. First, (see page 59) it proves to *you* that having had a heart attack does not mean that you're "crippled" for life. Patients used to think that once they'd suffered a heart attack, they were "finished"—in their work, their play, and their sexual activity. In the great majority of cases, this is pure nonsense. You'd be surprised how many of your tennis partners have cardiac histories. Then, too, a planned and supervised physical rehabilitation program improves physical fitness and cardiac efficiency. Several scientists, including Dr. John Longworth, at The University of California at Davis, have shown that animals, specifically pigs (whose hearts are very similar to our own), who exercise regularly develop a much more protective collateral circulation than those who do not. (Collateral circulation is discussed in more detail on page 20. It is an accessory circulation within the heart that takes over when the larger arteries close.)

We've come a long way in the forty years since I first started treating heart attack patients. In those early days we insisted on bed rest in the hospital for 6 weeks or longer! Today, you're up in a day or two because we now know that prolonged immobilization leaves you vulnerable to the formation of blood clots that can travel to your lungs and do you in. It also results in depression and lack of confidence in your future capabilities. Remember that *the great majority of heart attack patients who make it to the hospital do survive and go on to lead normal lives*. The key to their outlook is early diagnosis and modern treatment begun in time.

HEAT INJURIES: Just Cool It

Here's a statistic that may surprise you: *Very* hot weather causes more illness than any other natural phenomenon—more than cold, tornadoes, earthquakes, hurricanes, and floods. The best way to treat a heat injury depends on its severity, but every therapy includes cooling in some form.

The most common symptom of mild heat injury, either during or after physical effort under a hot sun, is *heat cramps,* usually in the legs. These cramps are due to the loss of salt and water from the body, and so may also be experienced in persons taking diuretics. The best treatment is lots of fluid with some salt added (one teaspoonful to every quart of water). Continue to be liberal with your salt intake for as long as your muscles are sore. Don't waste your money on commercial "replacement" formulas touted for this purpose. They're no better than plain salt and water.

The next grade of heat injury is *prostration* or *collapse,* obviously more disabling. Unlike simple heat cramping in which the body temperature is normal, heat prostration is accompanied by fever, nausea, vomiting, and dizziness. These symptoms are again due to loss of salt and water from the body. The best *treatment* for heat prostration is immediate transfer to a cool environment, and then lots of water with added salt. If you observe someone with heat prostration who is too sick or drowsy to care for himself, call an ambulance. In the hospital emergency room, he will be given the necessary fluids intravenously, observed for a few hours, and then usually sent home. Prompt replacement of lost fluid is especially important in the elderly and the chronically ill.

The most serious form of high temperature injury is *heat stroke,* a prime emergency usually associated with a variety of neurological symptoms ranging from confusion to coma. *Immediate hospitalization* is essential because death can occur within one hour after the brain has been damaged, literally cooked, by the excessive heat. Once in the hospital, the patient is quickly cooled by immersion in an ice bath and by the administration of intravenous fluids. Heart and lung function are constantly monitored to anticipate, prevent, and treat cardiovascular collapse.

Prevention is the key to dealing with heat-related disorders. Your body has several sophisticated mechanisms—sweating is one of them—to protect you against the ravages of heat excess. But these can be neutralized by certain drugs, which include antihistamines, tranquilizers, alcohol, and beta-blockers. So if your work or play requires intensive physical activity in very hot weather, use these agents either sparingly or not at all. Drink lots of fluid, wear loose clothing, and try to perform the bulk of your activities in the early morning or late afternoon and away from the blazing sun.

HEMORRHOIDS: Assessing Your Options

It's only a matter of time—you, too, will very probably develop hemorrhoids, if you haven't already. Four of every five Americans do. And you're in good company, because this condition has a long and distinguished history. Napoleon had so much pain (you know where) he wasn't able to sit on his horse when he should have started the Battle of Waterloo. The delay cost him the war. President Jimmy Carter went public with his hemorrhoids for no apparent reason. Who knows, this disclosure may have lost him the election. Hemorrhoids have caused so much suffering over the years that the Catholic church has designated Saint Fiacre as the patron saint to beseech for relief. (The patron saint of wine, on the other hand, is, believe it or not, Saint Isadore! I'm sure glad it wasn't the other way around!) If you have painful hemorrhoids, praying (in the standing position) won't hurt, but you should also follow the treatment I recommend below.

Hemorrhoids are rectal veins (part of everyone's anatomy) that have dilated, become inflamed, and developed clots. That may happen as a result of chronic constipation, recurrent diarrhea, stress, or multiple births. Persons who frequently lift heavy objects at work or at play, who sit a great deal without moving about, or are on their feet for long hours at a stretch, who are fat, don't eat enough fiber, or are simply getting older are also candidates for this disorder. You're vulnerable too if you are in the habit of reading gripping novels from which you simply can't tear yourself away while sitting on the toilet. (Always limit such intellectual pursuits in that location to 5 minutes per session.)

Hemorrhoids may be *external,* and protrude, or *internal,* the kind you can't see but can certainly feel. You can help prevent both varieties by eating lots of dietary fiber and drinking plenty of liquids.

Hemorrhoids that are just there and not causing any symptoms should be left alone. But if they're painful, bleed, itch, or protrude, here is what to do. Begin with a sitz bath in a tub with warm (not hot) water drawn to at least a 3- or 4-inch level, for 30 to 45 minutes twice a day. Adding some Epsom salts makes this siesta even more soothing. When you're through, apply a cream that contains a local anesthetic like dibucaine (Nupercainal). For additional relief, I advise my patients to insert a suppository that has a small amount of hydrocortisone (Anusol-HC) after each bowel movement or at least twice a day. These measures are most effective against *external* hemorrhoids, and will relieve symptoms of the occasional attack by reducing inflammation and swelling—until the next episode. But if you've begun to spend more time in the sitz bath than at the movies, or if the pain or bleeding become chronic despite all the suppositories and creams, then you'll have to submit to a direct attack on the hemorrhoids themselves.

What I recommend first for enlarged *internal* hemorrhoids is a simple procedure that can be done in a doctor's office. It involves tying the hemorrhoids with a small elastic band, after which they shrivel and shrink in a few days because their blood supply has been choked off (unless, of course, the rubber band slips off, which sometimes happens). If the hemorrhoids are too small for such banding, they can be injected with phenol or a similar substance that causes them to "sclerose"—that is, to scar and shrink. There are other methods to reduce the blood supply to the hemorrhoids—*freezing, infrared photocoagulation,* and *laser* procedures. Although these latter methods do the job, I'm not convinced that they are any less complicated or painful than banding or injection.

External hemorrhoids cannot be banded or injected. When they are huge and prolapsed so that they bulge out every time you move your bowels, they require actual surgical removal—a last resort.

One technique that is relatively new, and not yet widely used, and with which I have not yet had any experience, is coagulation by electrical stimulation. It is reported to be effective and relatively painless, with almost no side effects. None of the proctologists in my area are using electricity yet, but it is gaining popularity on the West Coast.

A word of caution. Don't assume that all rectal pain is due to hemorrhoids. You may have a *fissure,* a *skin tear,* or an *ulceration* in the anal area. If you do, bowel movements are a nightmare, especially when the stool is large and hard. Anal fissures also cause bleeding and soiling so that your underwear is always stained no matter how thoroughly you clean and wipe. If you have such a fissure, keep your stools soft by eating a high-fiber diet and drinking lots of water. Hemorrhoidal suppositories, creams, ointments, and a stool softener like docusate (Colace), 100 mg twice a day, will afford relief. If you're lucky, the fissure will heal, but surgical repair is often necessary. Make sure that whoever does the procedure is an expert in this particular field. A great ear, nose, and throat surgeon won't be of much help to you here. Unless properly fixed, anal fissures can become a chronic problem.

HEPATITIS: When It Doesn't Clear Up on Its Own

If you've been feeling out of sorts, have lost your appetite (especially for cigarettes), have a low-grade fever and a yellow hue to your skin and the whites of your eyes, you almost certainly have hepatitis—an infection of the liver. When you can remember that you ate raw shellfish a few weeks ago, or that someone with whom you had been in close contact in your jail cell, army barracks, or dorm at college had hepatitis, or you were vacationing in some exotic place where hygiene was not the strong suit, your hepatitis is most likely viral, type A. If, on the other hand, you are homosexual, promiscuous, or take drugs intravenously and share the needles, your hepatitis is more likely to be the B variety. Finally, if a few weeks ago you received blood transfusions, then I'd put my money on hepatitis C, which used to be called non-A, non-B. (The incidence of hepatitis C, a potentially serious infection, is decreasing dramatically now that the responsible virus has finally been isolated and can be screened out of blood for transfusion.) But your doctor will not rely on your memory alone to determine which type of hepatitis you have. He will send a specimen of your blood to the laboratory, looking for antibodies specific to A, B, or C. That's important, because what to do when you come down with the symptoms of viral hepatitis depends on which virus is infecting you. Each requires different intervention and carries with it its own prognosis.

There's no specific therapy for the *acute phase* of any form of viral hepatitis. In fact, it may be better not to take any medication at that time because the injured liver doesn't handle drugs very well. So whereas you might use a sedative or a pain killer to control the symptoms of most other illnesses, you probably should not do so with hepatitis. The best advice I can offer is to rest at home for a few days until your low-grade fever subsides, avoid alcohol, and then begin gradually to resume light activities. During this initial period, your doctor will be monitoring your liver function by means of blood tests. As soon as they begin to return *toward* normal, so can you. There's no reason for you to be hospitalized unless you develop some complication that requires treatment, like bleeding from the nose, mouth, rectum, or under the skin as a result of impaired liver function. In my experience, most patients with uncomplicated hepatitis of any variety are able to return to work full time in 3 or 4 weeks.

Doctors used to make a big fuss about diet in patients with hepatitis and until quite recently, routinely advised them to reduce their protein intake, eliminate fat, and abstain totally from alcohol. Only the last proscription is valid today because alcohol has a toxic effect on the liver. (As a matter of fact, it's a good idea to avoid all forms of alcohol for 3 or 4 months, or until your liver function tests have returned completely to normal.) But as far as diet is concerned, I permit my patients to eat whatever they wish. However, since their appetite isn't great to begin with, most instinctively reject fat and fried foods.

Should other members of your household avoid you like the plague? If you have hepatitis A (the food-borne variety), you are contagious for the first 2 weeks after exposure to the virus. Since there are usually no symptoms during this incubation period, if you think you were infected, avoid intimate contact during that interval. Hepatitis A is spread by the fecal-oral route, so wash your hands thoroughly after each bowel movement; make sure your bedsheets and other personal effects are not put into the washing machine with anyone else's laundry; and do not share your towels or toothbrushes with the rest of the family.

Most of these measures are unnecessary if you have hepatitis B or C, which are, for the most part, transmitted by blood. It's okay to move freely about the house, but do not share either your razor or your toothbrush, and abstain from any *intimate* contact during that period.

I'm often asked by someone who has been exposed to hepatitis A whether they need a protective "shot." If the contact has been really close (like living in the same household, or working together at adjacent desks), I recommend gamma globulin by injection as soon as possible. This enhances immunity temporarily and may either prevent the infection or render it less severe. But that doesn't mean rushing to have the gamma globulin just because you said hello to somebody on the bus who you later learned had hepatitis. If you are traveling to someplace where food and water may be contaminated, ask your doctor for gamma globulin just before you leave (unless you have already had hepatitis A, in which event you are immune to re-infection).

Gamma globulin is of no value against hepatitis B and C. If you prick yourself with a needle used by someone who may have hepatitis B, you should immediately obtain a shot of *hepatitis B immune globulin,* and get another a month later. Also, arrange to be vaccinated with the hepatitis B vaccine as soon as possible. All health workers and anyone else in contact with hepatitis B patients should be vaccinated. The immune globulin confers temporary immunity; the vaccine, on the other hand, protects for years by stimulating the formation of antibodies to the virus that destroy it.

To the best of my knowledge, hepatitis A infection in pregnant women does not cause congenital malformations in the fetus. But that's not true for hepatitis B and C. The fetus of any female so infected *is* at risk for developing serious hepatitis that can become chronic or even fatal. Whether or not to abort your pregnancy under these circumstances is something you should decide after discussing it with your husband and doctor.

Hepatitis A rarely if ever has any long-term consequences. However, hepatitis B and C are potentially serious. In about 10 percent of those infected, the hepatitis becomes chronic, damages the liver, and can eventually cause either cirrhosis (see page 69) or cancer of that organ.

Until recently there was nothing to be done for patients with chronic hepatitis B and C, but now treatment with recombinant *alfa interferon* (Intron-A) has been shown to have salutary effects in almost 50 percent of cases. It is expensive, and the exact dosage schedules are still being worked out, but prospects are very exciting. Steroids alone are ineffective in such cases, although pretreatment with them seems to enhance the action of the alfa interferon.

In summary, there is no treatment for the acute form of hepatitis other than good nutrition, lots of rest, and no alcohol. All cases of hepatitis A eventually clear completely, but 10 percent of hepatitis B and C become chronic. In these persons the most promising therapy at the moment is alfa interferon by injection.

HERPES: Out of the Limelight, But Still a Problem

In the pre-AIDS era of the 1970s, the focus among sexually transmitted diseases was on herpes. Remember some of the jokes circulating, like "Which one of the following *doesn't* belong in this grouping—gonorrhea, a good marriage, and herpes?" The correct answer was herpes, "because it's forever!" It's worth reviewing the latest treatment for herpes because this infection is still very much with us.

Whoever originated that herpes joke was right. Once you're infected, the virus remains with you forever and flares up unpredictably. When it does so in a woman near or during childbirth, her infant may die. In someone whose resistance or immunity is severely impaired by cancer, or antirejection drugs after organ transplantation, herpes can rage out of control and affect several body organ systems, occasionally fatally. But in the great majority of cases the sores that reappear from time to time, though unsightly, uncomfortable, embarrassing, and compromising to one's sexual schedule, are not a threat to life.

The hundreds of thousands of new cases of herpes each year are caused by two different viruses, HSV1 and HSV2 (herpes simplex viruses 1 and 2). HSV1 is the one you get from kissing, and is traditionally considered to have its home base in the mouth, while HSV2 erupts somewhere below the belly button. We used to think it was important to distinguish between them because HSV2 is more likely to result in recurrent infection. But there are so many variations in sexual technique and preference that either virus can be found where least expected. So from a practical point of view, the distinction between herpes 1 and 2 is not really worth bothering about.

In my book *Second Opinion,* written in 1982, I listed the therapeutic options then available for the treatment of herpes. None of them really worked. Today, however, we have acyclovir (Zovirax), which is extremely effective in minimizing the pain of herpes sores and decreasing the number of recurrences—but it is not a cure. The vaccine I was hoping for to *prevent* herpes when I wrote *Second Opinion* is not yet in sight.

Acyclovir (Zovirax), which comes in topical, oral, and intravenous formulations, does not actually destroy the herpes virus, but penetrates and interferes with the internal processes of the cells that harbor it, thus slowing the replication or growth of the virus. It is relatively nontoxic.

If you've been exposed to somebody with herpes and about 4 or 5 days later you develop burning, itching, or pain at the site of contact, immediately take 5 capsules a day of acyclovir (200 mg each) for 10 days. This will either abort the initial attack or shorten its duration and reduce its severity. Acyclovir is excreted by the kidney in a matter of hours, but a sick kidney cannot eliminate it efficiently, so take only half the dose if you have kidney trouble.

Since no one can predict when a *second* attack will occur, I advise my patients to stop the acyclovir after all symptoms from the initial one have cleared—and then wait. At the very first indication of recurrence (burning, itching, or pain), I have them resume the acyclovir immediately—the same 200 mg capsules, 5 times a day, but this time for only 5 days. That almost always minimizes the symptoms and shortens their duration. After *one* recurrence I wait to see how frequently any future eruptions will occur. If they do so every few months, I prescribe a maintenance dose of 2 acyclovir capsules twice a day. This prevents or significantly reduces the incidence of recurrence.

How long is it safe to keep taking acyclovir? Will your herpes virus eventually become resistant to it? As far as we know *now*, acyclovir is safe and effective for at least 3 full years. But it's more than likely that, as our experience with this antiviral agent continues, this "safe" period will turn out to be longer. However, at the present time, 3 years is all we can vouch for with certainty. Any patients who require Zovirax for longer than that should be examined every few months for evidence of toxicity.

There are other important considerations in the treatment of herpes. For example, most pregnant women who have been exposed to it should take acyclovir as a preventive measure to reduce the chances of their infant's becoming infected with the virus and suffering retarded growth, or even death. But that decision should be made by an infectious disease specialist, your gynecologist, and you—consulting together.

For severe generalized herpes infections in those with impaired resistance, the best treatment is *intravenous* acyclovir. If you are "immuno-compromised," but are not experiencing symptoms at present, it is very important for you to receive it anyway in order to reduce the risk of recurrence.

HICCUPS THAT JUST WON'T GO AWAY:

Not a Laughing Matter

When the nerve centers that control the muscles of respiration become irritated and go into spasm, you get the hiccups, and that can occur at the darnedest times. For example, I could never really share a good joke with my brother-in-law because every time he laughed, he developed hiccups that lasted for hours. After a while he wouldn't listen to my stories anymore; they just weren't worth it. Many years ago when I used a pipe, I'd often end up hiccuping after I had inhaled the warm smoke. Hot liquids do it to me still. Such "run of the mill" hiccups usually stop in less than an hour, but when they persist for days or even weeks, especially in the elderly and the sick, they need to be terminated. Hiccups may *sound* funny, but they sometimes mean trouble.

Ask ten different "experts" what to do about hiccups, and you'll get ten different answers. For example, my wife's method is to come up behind me silently and then loudly shout "boo" in my ear. I finally convinced her that hiccups are preferable to cardiac arrest. She now has me drink cold water and hold my breath. Another authority, my Colombian housekeeper, has a different technique. She dries leftover bread, reduces it to crumbs, and forces me to eat half a cup's worth. A clergyman I know advises members of his congregation to vomit to terminate an attack. There are doctors who will press on your eyeballs or neck; others even massage the rectum with the finger! I just stroke the patient's palate from side to side with a tongue depressor. All of these maneuvers have been known to work sometimes, but when they do, I'm not sure the hiccups would not have ended anyway—on their own!

When hiccups persist and are fatiguing, as, for example, after surgery, we may need drugs to terminate them. (But can you imagine doctors or nurses running up and down the ward shouting "boo," making people vomit, or passing out cups of crumbs?) The first agent to try is Compazine (prochlorperazine), 10 mg, 4 times a day. If that doesn't work within 24 hours, I recommend metoclopramide (Reglan), 10 mg, 3 or 4 times a day. Here's yet a third alternative. Researchers at the Walter Reed Army Medical Center reported in 1989 that nifedipine (Procardia, Adalat—a calcium channel blocker used mainly in the treatment of angina pectoris and high blood pressure) will stop hiccups in a dose of 10 mg, 3 times a day. I don't believe that anyone understands the mechanism(s) by which any of these agents work—if, in fact, they really do!

HIGH BLOOD PRESSURE: Even Though You're Calm, Collected, and Feeling Fine

Which one of the following statements is *always* correct? A hypertensive individual (a) has high blood pressure, (b) is nervous, (c) is tense. The right answer is "a," but you'd be surprised how many people would answer yes to all three—that anyone with high blood pressure must be "hyper," "nervous," and "tense." That's simply not true. While emotional stress may raise pressure temporarily, as, for example, when it's measured in a doctor's office ("white coat syndrome"), sustained elevation is a disease that has no predictable relationship to your personality.

It's too bad that *hypertension* is the medical term for high blood pressure because it just adds to the popular misconception that an elevated pressure will simply go away if you relax, "take it easy," or use tranquilizers. Of course, we should all try to deal with life's problems calmly and with equanimity. But if you suffer from true hypertension, relaxing may lower your blood pressure a little, but will not drop it down to normal. In order to do that, you will need to *lose weight, exercise, reduce your salt intake* (*in some cases*), *and if these measures don't work, take specific blood pressure lowering drugs.* It's not easy to get everyone to follow this advice because hypertension itself usually causes no symptoms until late in the course of the disease, while the drugs that control it often do. Unfortunately, however, if you fail to normalize your pressure, hypertension will give you more than side effects; it is very likely to kill you. The end may come in the form of a stroke, a heart attack, or be preceded by blindness, serious kidney disease, heart failure, or blocked arteries in your legs.

Because more people are now able to tolerate the drugs their doctors prescribe, there has been much greater and sustained compliance with treatment, a significant drop in the national blood pressure level, and an almost 50 percent reduction in the death rate from strokes in the last 15 years. The incidence of heart attacks has also decreased by 35 percent during the same interval, for which there may be other explanations in addition to better blood pressure control—less cholesterol and fats are being consumed, fewer people are smoking cigarettes, and more are watching their weight and exercising.

The fact that high blood pressure is a "silent killer" and a major cause of death and disability was not always appreciated. Medical books published in the early 1900s didn't even list hypertension in the index. And as recently as 1950, there were doctors who believed that as long as you were feeling fine, high blood pressure didn't mean a thing. The magnitude of the problem and its impact on health is now more realistically understood. Ten years ago we estimated that there were only 25 million Americans with hypertension. Today, as more and more "healthy" persons get their pressures checked routinely, that figure turns out to be closer to 60 million. In other words, one in every four or five of us in this country has an elevated blood pressure that could eventually do us in.

Despite all the time and money spent on hypertension research, we still don't understand why it happens in about 85 percent of cases. In 10 to 15 percent, a specific cause can be identified and usually involves kidney diseases, an overactive or underactive thyroid, or hormones secreted by the adrenal glands (adrenalin, noradrenalin, cortisol, aldosterone). Some of these abnormalities can be dramatically cured, often by surgery—e.g., removing a hormone-secreting tumor, or opening up a narrowed artery to the kidneys by balloon angioplasty or an operation. Your physician will know when and how to look for these relatively rare conditions.

But even in the majority of cases in which the doctor does not know precisely *why* your pressure is too high, it can now easily be normalized—and that's what's important. Not so long ago, side effects from the handful of drugs then available to treat hypertension were often intolerable. Many patients preferred to take their chances with the late manifestations of the disease rather than suffer the symptoms resulting from its treatment.

But in the past decade, literally hundreds of new medications that are much easier to take have become available for the management of high blood pressure. There's no longer any reason for you to fear their consequences. If you can't tolerate a particular one, there are several others on which to fall back. Effective treatment no longer means sacrificing potency (if you're a man) or sexual desire (if you're a woman); dizziness, dry mouth, weakness, fatigue, and running to the john all night can also be avoided.

There are several important misconceptions about hypertension treatment. Take the matter of salt, for example. It's not true that everyone with hypertension must follow a salt-free diet. Only *certain* individuals need to reduce their intake appreciably. You can determine if you're one of them by going on a strict low-salt diet for a few weeks. If your blood pressure doesn't drop much, then you are probably not salt-sensitive and can enjoy the health advantages (and pleasures) of moderate salt intake without risk.

In my opinion, too much alcohol probably contributes more to elevated blood pressure than does excessive salt. So if your doctor recommends you follow a rigid low-sodium (salt) diet, ask him whether he knows *for sure* that it's important in *your* case. Life is tough enough without the additional burden of a drab diet, especially if it's unnecessary. But, if you're one of those for whom salt is truly bad, there are some pretty tasty herbal combinations and salt substitutes you can use (see page 254).

Doctors treating hypertension used to aim for the lowest possible pressure that could be tolerated without dizziness or profound weakness. They and their patients prided themselves on achieving such low levels, on the assumption that the lower the pressure, the fewer the late complications of hypertension. I don't think that's true. *Too* low a pressure, even one with which you feel well, can be dangerous especially if you're elderly, or have some form of heart disease or a past history of a stroke.

If I've convinced you that detecting and treating high blood pressure is no big deal yet extremely important (even if you're calm, collected, and feeling fine), *arrange to have your pressure checked.* If it's normal, congratulations! But what's normal, and how high is high? In my own practice, I begin to "keep tabs" on (but do not treat) healthy subjects with readings of 145/90. Doctors do not always agree about when to start therapy. I usually do so when the numbers exceed 155/90 on at least three successive visits. However, in the presence of co-existing disease of the heart, eyes, or kidneys, and such other risk factors as smoking or high blood sugar or abnormal lipids, I am inclined to begin treatment at lower levels. None of these conditions cause high blood pressure (except certain forms of kidney trouble), but they are aggravated by hypertension.

Let's assume that you and your doctor have agreed that your pressure should be lowered. What's the best way to do it? Unless the reading is *very* high—like 200 or more on the top side, and greater than 110 for the bottom figure—don't rush to take pills.

First make sure you're not on any preparation that itself can *elevate* your pressure! For example, phenylpropanolamine (PPA), an ingredient in many cold remedies, decongestants, and appetite suppressants, can raise your pressure when taken in excess, or in combination with coffee. So will cortisone used over the long term for whatever reason—arthritis, asthma, or cancer.

If there is no drug causing your high blood pressure, and you're overweight, give yourself about 3 months of dedicated dieting before starting antihypertensive medication. If you can shed the excess poundage in that interval, there's a good chance your blood pressure will come down without any other measures. This is also the time for you to experiment with a low-salt diet—one with which you can live—to see whether it really makes a difference in *your* case. The basic rules here are to avoid salty foods like herring, lox (the bagels with cream cheese are okay), and bacon. Don't even *think* of putting a salt shaker on the table, let alone using it. You may, however, add a small amount of salt in the cooking to make your food more palatable. If your blood pressure doesn't fall to acceptable levels after 3 months on a low-calorie, low-salt diet, then you may as well resign yourself to medication.

But remember, nothing is forever. Although most antihistamine drugs are usually continued for a lifetime, I often stop them after a year to see if the patient can do without them. Not infrequently they can. And even after drugs have been prescribed, don't give up trying to lose weight, because the thinner you are, the fewer pills you'll need. Side effects from any drug, no matter how slight, are dose-related. The less you take of any medication, the better you'll feel. It's also a good idea to join an exercise program (if you've been cleared by your doctor to do so). Although exercise alone won't lower your blood pressure significantly—neither will "relaxation" techniques, at least not in my experience—all these steps, when taken together, make medication work more effectively and in smaller doses.

Drugs for Lowering Blood Pressure

We come now to the nitty-gritty of treatment—the drugs. There are no fewer than 200 prescription medicines available to reduce blood pressure. However, they all fall into only five major drug categories, each of which lowers blood pressure in a different way: the *diuretics,* the *beta-blockers,* the *calcium channel blockers,* the *ACE inhibitors,* and drugs that relax the increased tone or *spasm* of the arterial walls that accompanies hypertension. Several preparations combine two or more different agents.

Before starting any blood pressure medication, always ask your doctor about possible side effects—impotence, diarrhea, constipation, frequent urination, fatigue, a rash, dizziness, worsening of asthma, a slow pulse, and so on. But don't let this litany scare you off. It's all *potential,* not *inevitable,* and very much a matter of individual response. One person will tolerate a drug in large amounts, while another reacts badly to a tiny dose. Also, if you're being treated by more than one doctor because you have several different "conditions," make sure each of them knows what medications the others are prescribing. Drug interaction during in the treatment of hypertension can be dangerous.

Unlike a cough medicine or a pain killer, the best drug with which to treat *your* high blood pressure depends on such factors as your age, your sex, your general state of health, what other ailment(s) affect you, and yes, even your race. So I can't make a blanket recommendation that will apply to everyone. I normally have my Caucasian (white) patients who are otherwise healthy except for their hypertension, start with a small dose of an ACE inhibitor. ACE stands for *a*ngiotensin *c*onverting *e*nzyme. Angiotensin is a natural substance in everyone's body that raises blood pressure by constricting the blood vessels. An ACE inhibitor "inhibits" that action. I prefer ACE inhibitors because in my experience, they have the fewest side effects. For example, unlike diuretics, they don't raise your cholesterol or make you spend the better part of your life in the bathroom emptying your bladder; unlike beta-blockers, they don't slow you down, give you bad dreams, worsen your asthma, induce heart failure, or drive you to reading porn magazines to help restore some interest in sex. On the down side, the ACE inhibitors cost more than diuretics and beta-blockers, and there are no generic forms as yet.

There is another good reason to start with an ACE inhibitor. If it results in a dramatic drop in pressure, then you know that renin is an important factor in the causation of your hypertension. (Renin is the blood pressure–raising hormone made by the kidneys that is blocked by the ACE inhibitor.) Excessive amounts of renin are produced when the arteries going to the kidneys are narrowed. So if you respond to an ACE inhibitor, it may be worth looking at those arteries to see if they can be dilated by balloon angioplasty. On the other hand, when an ACE inhibitor is not effective, that more or less rules out renin and suggests instead a salt-dependent type of hypertension for which calcium antagonists or diuretics are more effective. There are currently four ACE inhibitors available in the United States: Captopril (Capoten), enalapril (Vasotec), lisinopril (Prinivil, Zestril), and ramapril (Altace). Although all four lower blood pressure in the same way, they are not altogether interchangeable. Captopril acts the fastest, so you can make a quicker judgment about its usefulness, but it is also short-acting. Zestril is the most potent, and its duration of action is the longest, so that it can often be given only once a day. For long-term treatment of hypertension, I usually have my patients take 10 mg of Zestril before breakfast. We recheck their pressure in about 7 to 10 days (I say "we" because I encourage my patients to record their own blood pressure at home). The side effects to look for are:

- *A cough.* I don't think anyone knows why it happens, but all ACE inhibitors cause a cough in about 20 percent of patients after a few weeks. You should suspect this drug if the cough is "dry" and there are no other symptoms of a cold or fever. Unless you're aware of this side effect, you may end up going from one specialist to another—having your sinuses drained, your vocal cords checked, your lungs x-rayed and re-x-rayed, until someone just happens to remember that ACE inhibitors may induce cough. It will disappear shortly after you stop the medication.

- *A rash,* which clears up when the drug is withheld. As a matter of fact, it may vanish even if you continue the therapy.
- *Dizziness and faintness,* especially when blood pressure falls too far and too fast.
- *Exhaustion.* Patients are often worn out when the blood pressure is reduced from a previously high level, and unfairly blame it on the drug. However, after the body becomes accustomed to the new pressure levels, energy is usually restored.
- *Elevated potassium,* so do *not* use potassium supplements with ACE inhibitors (and remember that many salt substitutes contain potassium).

Do not take an ACE inhibitor if (1) you have *kidney disease,* (2) you're *pregnant* (no one really knows the effect of these drugs on the fetus), (3) you're *breast feeding,* or (4) you have severe *narrowing of the aortic valve in the heart.*

If a 10 mg Zestril capsule doesn't do the job, I add a small amount of a diuretic like hydrochlorothiazide (HydroDIURIL), 25 mg a day. This combination is effective and well tolerated. If it is not, I then switch to another category of drugs, namely calcium channel blockers, rather than increasing the dose of the Zestril and/or the diuretic.

Even though I prefer to treat my patients with ACE inhibitors, many do better with different drugs. For example, African-Americans with high blood pressure respond best to diuretics so that's what I prescribe for them first. However, before administering any drug to black patients with hypertension, I have them follow a salt-restricted diet because elevated pressure in these individuals is usually "salt-sensitive." It's only after diet alone proves to be ineffective that I start one of the many diuretic preparations available.

Diuretics are the antihypertensive agents your doctor is most likely to prescribe first. These were among the earliest drugs available for this purpose, and since familiarity breeds inertia, many physicians are most comfortable with them. Diuretics do have some advantages. They are the least expensive of the blood pressure medications, and they are especially effective in blacks and in persons with low renin values. But except in such patients I do not prescribe them very often, for the following reasons. Statistically speaking, when blood pressure has been lowered *by a diuretic,* the incidence of heart attacks is not decreased. We used to think that as long as your blood pressure stayed down, no matter what medication was used, the risk of heart attack was reduced. That appears to be true for almost every antihypertensive drug *except* the diuretics. In other words, when blood pressure has been normalized by using diuretics, the incidence of stroke is decreased, but *not* the incidence of heart attacks. That may be so because while they do reduce the blood pressure, diuretics also act adversely by raising cholesterol and lowering HDL levels (the good cholesterol fraction). There are other disadvantages to the diuretics as well. Although they decrease the blood volume (and thus the blood pressure) by increasing the elimination of salt and water by the body, they also result in the loss of other valuable minerals including potassium and magnesium. Deficiency of these two substances leaves you weak and tired, and low potassium can induce a serious disturbance of heart rhythm, especially if you are taking some form of digitalis. One man in three who takes diuretics becomes impotent. These agents often raise blood sugar (which, however, returns to normal when you stop the drug). They elevate uric acid levels in the blood, and a high uric acid causes gout (see page 114) so that if you're vulnerable, you may wake up one morning with a big, red, swollen, painful toe. They can upset an empty stomach (so always have your diuretic with food or milk). And then there's the matter of voiding. Diuretics are usually taken once a day, either with breakfast or at bedtime. A morning dose will keep you "running" all day; taken at bedtime you're likely to get up every hour or so during the night to empty your bladder.

For all these reasons, I prefer not to prescribe diuretics as first-line drugs in white hypertensive patients. However, despite their drawbacks as antihypertensives, they do have an important place in the treatment of heart failure, whenever the body retains fluids and in certain "sodium-sensitive" forms of high blood pressure, especially common in blacks.

Calcium channel blockers are good antihypertensive drugs too. They are also very effective against angina, so that if you're "lucky" enough to have both angina pectoris and high blood pressure, a calcium channel blocker fits the bill. Those currently on the market in the United States, with more on the way, are nifedipine (Procardia, Adalat), nicardipine (Cardene), verapamil (Isoptin, Calan), diltiazem (Cardizem), and isradipine (DynaCirc). Although they all lower blood pressure by dilating the arteries, these agents are not interchangeable because each has certain specific characteristics that may be good for one person and not for another. Their side effects include constipation, dry mouth, nausea, cramps, and in rare instances impotence. Procardia (nifedipine) and related drugs are especially likely to cause heart palpitations and swelling of one or both feet and ankles. Constipation is more of a problem with verapamil. Like any other drug, any of these agents can cause a rash.

Beta-blockers, which relax the blood vessels by blocking certain nerve impulses, are very versatile drugs. These agents have some twenty different approved uses or "indications." Besides lowering blood pressure, they prevent anxiety and panic attacks, a boon for entertainers and public speakers; they are a mainstay in the treatment of angina, irregular heartbeats, and heart attacks; they control tremors; they make withdrawal from alcohol more tolerable; they're effective in preventing migraine headaches; two of them, timolol (Timoptic) and betaxolol (Betoptic), are widely used to reduce elevated pressure in the eye due to glaucoma; they are used in thyroid disorders; one is even a vaginal contraceptive. However, they should be avoided if you have chronic lung disease like emphysema, bronchitis, or asthma because they can induce bronchial spasm; or they can worsen matters if your heart rate is slow to begin with. They often cause impotence, fatigue, bad dreams, and constipation but many of these side effects are dose-related; also if one particular brand is intolerable, another may work.

Prazosin (Minipress) and its newer, longer acting relatives, terazosin (Hytrin) and doxazosin (Cardura), reduce blood pressure by specifically blocking nerve receptors in the *artery walls*, thereby relaxing and widening these vessels. The very first time you take them, they can make you dizzy and even cause you to faint, especially when you stand up suddenly from the lying position. So always start treatment with any of these agents at bedtime, and let your doctor know if they leave you lightheaded or feeling faint before you take a second dose. Both Minipress and Hytrin do confer one important bonus for men whose prostate is enlarged: They often reduce the number of times the bladder needs to be emptied at night.

Clonidine (Catapres), which comes both in oral and transdermal formulations (the latter is absorbed from a patch on the skin), lowers blood pressure by dilating the blood vessels. It acts on the brain in such a way as to block nervous signals to the arteries, which may put them into spasm. This drug is also useful in controlling symptoms of drug, alcohol, and tobacco withdrawal. It's not my favorite agent for high blood pressure because patients seem to develop "tolerance" to it so that more and more of the drug is required to keep the blood pressure down. And when stopped suddenly, it may result in a severe rebound elevation of the pressure.

One medication I absolutely shun is reserpine. It was widely used years ago when there was little else available to reduce blood pressure. However, it also affects nerves in the brain, and causes severe depression that can be prolonged and life-threatening, especially in older persons.

Given the wide variety of agents available for treating high blood pressure, you should have no trouble in getting yours under control. Combining small doses of virtually all these drugs can lower blood pressure effectively and safely. There's no need for anyone to be cooking a stroke or a heart attack these days. And remember, don't suffer in silence. If you have lived with drug side effects over the years in the mistaken belief that that's the price you have to pay in order to avoid the complications of hypertension, ask your doctor to take a fresh look at what you're taking. But the initiative must come from you. Doctors are concerned with the bottom line; they don't like to rock the boat when the blood pressure readings look good, even if *you* feel lousy. But good numbers do not require disabling toxicity.

Every patient with hypertension should know that while treatment seems to be focused on normalizing the numbers, *the real goal in managing high blood pressure is prolongation of useful life by preventing heart attacks and strokes.* That requires not only control of blood pressure, but regulating associated phenomena such as diabetes, tobacco, cholesterol, obesity, and other less well defined factors like stress.

IMPOTENCE: Raising Expectations

According to statistics (how they were obtained, and by whom, I'll never know), 20 percent of men are not able to achieve an erection by the time they reach age sixty, although another 20 percent are still potent at age ninety. In my office experience, the 20 percent figure for the sixty-year-olds is too low, and the figure for the nonogenarians is, for the most part, wishful thinking. But for those who still "can" at that age, may the force be with them!

Part of the problem in obtaining accurate data on so private a subject as erection is its definition. What fills some men with pride is laughable to some women. Also, most of us don't brag about our inadequacy. Although more males are coming clean to the doctor with respect to sexual problems, largely because they believe, sometimes correctly, that the medications they have been given are to blame, impotence is still very much in the closet.

Here are the biological requirements for a noteworthy erection at any age. Assuming you have been adequately stimulated, your body needs to be making enough *testosterone* (male hormone) to start the process. If the testes, which produce testosterone, have been surgically removed (as is often the case in prostate cancer), or are no longer working as well because of age, then even the most intense sexual provocation will do nothing more than leave you wondering what all the fuss is about.

If you do have a bountiful supply of testosterone, and have been properly titillated, you need an *intact* brain to appreciate the good news. A hot sex message received by a healthy brain is then sent along nervous pathways to the genital area where it is translated into the payoff. Any disease or injury that interferes with that transmission will leave you excited up top and flaccid down below.

The third requirement for an adequate erection, in addition to sufficient testosterone and an intact nervous system, is a circulation that can deliver the extra blood needed to stiffen the penis. Arteriosclerosis of the vascular system, which blocks the arteries to the genitals, makes such erection either feeble or impossible.

Finally, any injury to the pelvic contents as, for example, extensive surgery of the prostate gland for cancer (in which the nerves supplying the penis may be severed), or a major abdominal operation (when the blood supply and/or nerves in the area are compromised), can result in impotence, too.

Even when all the components necessary to ensure an erection are present, you may still not enjoy one if there is either a *psychological* block to normal sexual activity, or you're taking an impotence-inducing agent such as an antihypertensive drug, a tranquilizer, a sleeping pill, or a narcotic or are drinking alcohol to excess.

The best treatment for impotence first requires identifying and then eliminating its specific cause—psychological, pharmacological, hormonal, neurological, or vascular. But the exciting news in this field is that even when such factors have been identified but are themselves not "treatable," there are new ways that make it possible for most men to enjoy sexual activity.

Psychological causes of impotence usually develop as a result of some specific anxiety-producing event, as, for example, a recent heart attack, the fear of contracting a sexually transmitted disease, loss of interest in one's long-time sexual partner, depression, overwork, fatigue, business worries—whatever. In these circumstances, counseling, reassurance, and review of sexual techniques and expectations may be helpful.

When the problem is *vascular* due to blockage of blood vessels to the penis, surgery can sometimes cure the problem. But, when impotence is due to *neurological* conditions such as multiple sclerosis or nerve injury associated with diseases like diabetes or a stroke, less can be done. When *testos-*

terone deficiency is the main culprit, its replenishment may solve the problem. However, this doesn't occur as often as you might think. Many men ask their doctors for hormone "shots" when their testosterone levels are perfectly adequate, and there are other reasons for their impotence.

If a psychiatrist can't help you, or you're not a candidate for vascular surgery, if the nerves to the penis are shot and beyond repair, if you don't have a hormone problem, and there's no specific drug that's causing the impotence, then there are several mechanical aids that can help. But first try *yohimbine,* derived from the bark of the Yohimb tree in South America. The Indians there claim that boiling it and drinking its broth or eating it increase libido and/or actual sexual performance. Yohimb, the active ingredient in the tree bark in its pure form, does, in fact, increase blood flow to the penis and reduce the exit of blood from it—an ideal combination as far as erections are concerned. Some reports in the medical literature attest to an almost 50 percent response rate after 2 to 3 weeks of therapy with yohimbine in men rendered impotent either by diabetes or psychological factors. Personally, I have seen it "work" only once or twice. There are several brand name preparations of yohimbine, but I recommend the generic form. It's cheaper than, and just as effective as, the brand names. It is available only by prescription and comes in 5.4 mg tablets. The suggested dose is 3 tablets a day. But don't use yohimbine if you're on tranquilizers or sleeping pills, or have any kind of kidney problem. As far as side effects are concerned, it may elevate the blood pressure, speed up your heart rate, and leave you feeling nervous. If that happens, reduce the dosage by half and then gradually increase it to the 3 a day dose.

If the yohimbine doesn't work, ask your urologist about a *vacuum constrictor device* (VCD), which may well solve your problem. The only one with which my patients have any experience is Erec-Aid, made by the Osbon Company. It costs about $350, and I'm told they'll return your money within 60 days if you're not completely satisfied (your partner's satisfaction is not part of the deal). The Osbon unit consists of a lightweight plastic vacuum cylinder, a small pump, and some silicone rings. The penis is inserted into the cylinder and the hand-operated pump creates a vacuum that results in blood flow into the penis, mimicking what nature itself does. When enough blood has entered the organ and left it rigid, one or more rings are slipped off the cylinder and onto the base of the penis to keep the blood from leaving the organ. This permits the erection to be maintained for as long as 30 minutes even after orgasm. The Osbon unit is reported to work in more than 90 percent of cases, and is safe. Occasionally, there is bruising, and the erection itself may not be as impressive as it was 30 years ago. But try it; I'll bet you'll love it!

If the vacuum constrictor device is not to your liking, there is another effective option—*self-injection* of a mixture of two drugs, papaverine and phentolamine, into the penis itself. This method is very much preferred to the Osbon by the "erectile dysfunction" team at my own hospital, but some of my patients don't agree. However, the injections won't work if you have severe vascular disease (neither will the Osbon), or in most diabetics older than sixty. My own patients have complained of bruising when a vein is inadvertently punctured, or pain and tingling when the needle hits a nerve. Some men develop temporary liver abnormalities from the medication, while in others, little nodules form in the penile tissue, probably because of improper injection techniques. In some individuals, too, the erection doesn't go down for hours, which is not as wonderful as you might think since the pain exceeds the pleasure! But overall, this is an important and effective development in the treatment of impotence. The erection lasts considerably longer than with the vacuum device—1 to 2 hours, during which time you can enjoy sexual intercourse to your heart's content. But because the long-term effects of these injections into the penis are not really known, I suggest that you limit yourself to 2 shots per week. Play golf the rest of the time.

I used to be enthusiastic about *penile prostheses,* but I no longer recommend them. Rods, which keep the penis stiff, are surgically and permanently inserted into its shaft. There are three types—rigid, semirigid, and inflatable. With any of these, you lose the ability to have a spontaneous erection in the future, because the rods destroy the penile tissue. So never agree to such a prosthesis if your problem is temporary or recent. There are other complications from these procedures as well—infection, scarring, and malfunction. So although prostheses often do work, I would never opt for one myself.

There is one final therapy for impotence you may wish to consider. The section on angina (page 19) describes the artery-dilating properties of nitroglycerin, a drug widely used to "open up" the coronary arteries in the heart. According to some researchers, nitro has the same effect on the penis. If impotence is a matter of insufficient blood flow to that organ, why not, they reasoned, coat it with the transdermal nitroglycerin we give to persons with angina. They did, and guess what happened? It worked! There was sufficient engorgement to permit vaginal penetration! But, quite seriously, if you're going to try this approach, use a condom. When nitroglycerin is introduced into the vagina and absorbed, it may give your partner a headache!

INCONTINENCE: When You Can't Control Your Urine

Not long ago, I sat down in my examining room to chat with my last patient of the day. My chair is like a movie director's, with a canvas seat. Over the years with continued use (and my increasing weight) it has begun to "give" so that it is now somewhat concave. Soon after I settled into it, I became aware of a cold, wet feeling spreading over my buttocks, such as you might expect when dipping your bottom into cool water. I stood up very slowly—and looked. To my surprise, the back of my white coat was literally dripping with a yellow liquid that looked very much like urine! If *I* hadn't done it, who had? Then I remembered the previous patient, and the mystery was solved. She was an elderly lady who had recently had a stroke and who was now unable to control her urine. She had sat in that chair waiting for me, during which time she had "let go" (involuntarily, of course), creating a huge puddle in my concave director's chair.

Some 20 percent of older persons living at home, as well as a third of those in hospitals and half of all nursing home residents, are incontinent of urine. Although incontinence is an enormous problem among the elderly, it can affect persons of any age. It is not only embarrassing and inconvenient, it often results in skin and urinary tract infections.

Incontinence is usually referred to as *bedwetting* in children, *stress incontinence* in women, and *spontaneous incontinence* in the elderly of both sexes. It may never come to light unless the doctor specifically asks about it or notes wet panties or urine-stained underwear. It is often considered to be a "normal complication" of aging (which it is not) or ignored for fear that it may lead to surgery. Many persons prefer to go through life wearing diapers than have an operation. The fact is, incontinence *can* often be dramatically improved and sometimes even cured, without surgery. So don't keep it a secret—at least not from your doctor!

In order to understand the treatment of incontinence you need to understand the normal urinary mechanisms. The urinary tract consists of the *bladder* (where the urine made by the kidney is stored until it is ready to be voided) and the *urethra* (the passage through which it flows from the bladder to leave the body). In order for urine not to leak, the pressure in the urethra must be greater than that in the bladder. If it weren't, you'd constantly be losing urine. In a normal individual who is ready to "go" when the bladder is full, the muscles and nerves of the urethra are *voluntarily* relaxed so that the pressure within it drops and the urine flows out. After voiding, the pressure in the urethra rises once more and stays up until the next time. However, when the pressure in the bladder is chronically higher than that in the urethra, as is the case when the nerves and/or muscles controlling the urethra have been injured (in women who have had multiple pregnancies, in men after prostate surgery, and in certain neurological conditions like stroke, Alzheimer's disease, or Parkinson's disease), so that nerve signals indicating a full bladder are not transmitted, the result is incontinence.

The best treatment for this type of incontinence is the avoidance of anything that can exaggerate the impairment of the muscular or nervous control of the bladder and urethra—too much alcohol or coffee, cigarettes (if they make you cough), diuretics (water pills), the beta-blockers, various antispasmodics, antidepressants, antihistamines, cold medications, and even some of the drugs used in the treatment of asthma such as Ventolin (albuterol).

By far the most common type of incontinence is *stress incontinence,* which affects women much more than it does men. It can be caused by any of the conditions described below. Coughing, sneezing, laughing very hard, vigorous sports, sexual intercourse, all give rise to a temporary increase in abdominal pressure that is transmitted to the bladder. This results in an involuntary release of varying amounts of urine—a few drops to less than a teaspoon. So by definition, stress incontinence is *always* provoked by some specific activity.

In *postmenopausal women,* the lack of estrogen causes the vaginal walls to dry up and shrink so that they provide less support for the urethra. The result is a drop in urethral pressure and urine flows through it more easily—too easily, in fact. So if you've become incontinent after the menopause, ask your gynecologist to check your estrogen level (easily done by looking at the vaginal cells under the microscope). If it's too low, Premarin vaginal cream (conjugated estrogens) taken intravaginally, 2 to 4 grams a day, 3 weeks on, 1 week off, will usually solve the problem without side effects.

Incontinence in women is sometimes also caused by a *low-grade urinary tract infection.* When this can be eradicated by the appropriate antibiotic, the problem is solved.

When the *tissues around the urethra are lax* because they've been stretched by multiple deliveries, a fitted pessary (a diaphragm-like device introduced vaginally) often provides enough support to control the resulting incontinence. A newer treatment consists of the injection of collagen around the urethra, but this has not yet stood the test of time. If nonsurgical approaches fail, an operation that firms up or supports the tissues in and about the uterus and vagina will stop the incontinence in about 90 percent of cases.

If you wet your undergarments when you stand up after sitting for a while, or when you hear the sound of running water, or you have to void several times both during the night *and* the day, or you're able to "hold" it for only a few seconds after you get the urge and don't always reach the bathroom in time, chances are you've got some *bladder problem*. Either it's contracting independently of your wishes, perhaps as a result of a stroke, or the pressure within it is greater than that in the urethra. Here is the best way to remedy that problem. Practice holding your urine as long as you can. That strengthens the muscles in the area. After you can control it comfortably for one hour, add 15 minutes a week until you can wait for as long as 3 hours without distress. Also, take Tofranil (imipramine), 10 to 20 mg, 3 times a day for a few weeks then stop. The disorder may or may not recur. If it does, you can re-start it. Although Tofranil is commonly used as an antidepressant, when used in the treatment of incontinence due to bladder malfunction it acts on the muscles and nerves. If it gives you a dry mouth, chew some gum or suck on a lozenge, but persevere.

In some men, a *large prostate* may obstruct the urethral duct so that the urine can't get through it easily. The bladder then continues to fill, the urine overflows, causing incontinence or dribbling. The ultimate treatment for this situation is removal of the prostate gland.

If you've become incontinent *after prostate surgery,* don't be discouraged. Normal voiding is usually restored in a few weeks. When it isn't, drugs may help if the problem is mild.

When incontinence does not respond to medical or surgical therapy, there are special "diapers" you can buy that will hold almost half a liter of urine without your skin getting at all wet. You won't have any trouble identifying them on the drugstore shelf. These garments are particularly useful in patients with Alzheimer's disease. Meticulous hygiene is extremely important for such persons in order to avoid bed sores due to chronic dampness of the genital areas and buttocks.

Incontinence in Children—Bedwetting

Don't worry about bedwetting in children before the age of five. Beyond that, consistent bedwetting calls for a complete urological evaluation to see if there is some underlying physical or neurological problem. Whatever you do, never punish a kid who wets, and don't withhold water at bedtime either, as some well-intentioned parents often do. None of this helps one bit.

We used to think that bedwetting was largely psychological and due to harsh toilet training. That's not the case, although stress does worsen matters. The best way to deal with bedwetting is to reassure, encourage, and show your child that you have confidence in him or her. Forget about wiring them with alarms and other gadgets that only frighten the living daylights out of them. I also suggest using Tofranil (imipramine) in a sustained-release form (Tofranil PM), 25 mg, about an hour before bedtime in children between the ages of six and eight, and twice that dose in older ones. Continue this treatment for at least 2 or 3 weeks, and if it helps, then maintain it for 4 or 5 months. That often solves the problem permanently. When you've decided to discontinue Tofranil, don't do it abruptly. Decrease the dose gradually over a 3- or 4-week period so as not to induce any symptoms of withdrawal.

If your child is still bedwetting after age six, and nothing else works, there is a *new* breakthrough medication that may solve the problem. It's called desmopressin (DDAVP Nasal Spray), and the dosage is 2 nasal puffs a day for 3 months. But you mustn't miss a single day! More than 70 percent of the kids achieve "dryness" with this treatment, but not all remain controlled after the treatment is stopped.

Interestingly, the single most predictable factor in bedwetting children is the family history. When both parents themselves were bedwetters, the chances of their offspring doing so too are extremely high. When neither parent had the problem, none of their children is likely to have it beyond the age of five.

If you are worried about the continued incontinence in your child, here are some statistics that should cheer you up. Twenty percent of children wet their beds at age five; only 5 percent do so at age ten, and the figure drops to 2 percent at age fifteen. One in every 100 childhood bedwetters continues to have a problem for the rest of his or her life.

INDIGESTION: The Great Mimic

Years ago, lots of people died from "indigestion" because they didn't realize they were having a heart attack. They took an antacid instead of going to the hospital. But the pendulum has now swung in the other direction. There has been so much emphasis on the fact that indigestion often mimics a coronary that many people attribute *every* gas pain, with and without a belch, to the heart. Before I tell you what to do for real indigestion, let's make sure you've got your diagnostic signals straight. Here are the key points to remember:

- The heart is *not* situated in the left side of the chest, but dead center behind the breast bone.
- Cardiac pain is almost always dull and oppressive, like an elephant sitting on your chest. It is rarely sharp or sticking.
- A heart attack is usually accompanied by a cold sweat. Indigestion never is!

Once you've determined that the pain, discomfort, or fullness in the chest after eating a heavy meal is not from the heart, that doesn't automatically mean indigestion. You must still consider hiatus hernia (protrusion of the stomach up into the chest cavity so that its acid refluxes into the food pipe), gallbladder disease, spasm of the esophagus (food pipe), or peptic ulcer. When all the appropriate tests have been done, only one person in five will turn out to have true indigestion. Some other condition accounts for the rest.

The real cause of indigestion remains a mystery. I suspect it's probably the result of abnormal contractions or malfunction of the food pipe rather than too much acid. That's why antacids and drugs that block the formation of acid by the stomach such as Tagamet (cimetidine) and Zantac (ranitidine) don't work. Neither will pancreatic enzyme substitutes or antigas preparations like simethicone (Gaviscon). A good belch *will* help, at least temporarily. And if the attacks of indigestion are frequent and troublesome, try a 20 mg capsule of Prilosec (omeprazole) in the morning.

In summary, chronic "indigestion" is basically a wastebasket diagnosis, but one that may mimic several potentially serious disorders. Once these have been excluded after a careful work up, and you're left with "indigestion," resign yourself to the fact that there is neither a cure nor a really effective treatment. A good belch will go a long way!

INFECTIOUS MONONUCLEOSIS AND THE CHRONIC FATIGUE SYNDROME:

Missing—A Specific Treatment

Infectious mononucleosis ("mono" or the "kissing disease," as it is sometimes affectionately called) is caused by the Epstein-Barr virus (EBV), and generally affects the young. (I can't remember an active case in anyone over the age of forty in my own practice.) Ninety percent of Americans have been so infected, and their blood contains the antibodies to prove it, even though their "mono" may not have been recognized as such at the time. The infection usually mimics a mild cold with a low-grade fever and a slightly sore throat. But it can also present itself as a *bad* sore throat with substantial fever and obvious glandular enlargement, especially in the neck. These symptoms generally persist for a few weeks and then disappear. Occasionally a patient will feel washed-out for weeks or months on end. No one knows why this happens, how to predict it, how to prevent it, or how to treat it. Theoretically, one attack should provide lifelong immunity, but from time to time I see cases in which the disease seems to have recurred.

There is a great temptation, because of the sore throat, the tender glands, and the fever, to take an antibiotic when you have "mono." Don't do it! It's either useless or, as is the case with ampicillin, may cause a widespread rash and leave you sicker than you were. The best treatment for proven mono is nothing more than a few days of rest, although there's no proof that going about your business if you have the energy to do so really makes any difference one way or another. You should, of course, avoid intimate (kissing) contact until your symptoms have cleared.

If you have mono, do not engage in any contact sport—in the unlikely event that you feel like playing any games! *Roughhousing can rupture the spleen.* This organ, situated in the upper left side of your abdomen, enlarges and becomes soft in persons with this disease. Since it lies close to the surface of the body, it is vulnerable to injury. That's why doctors don't poke the belly too vigorously when examining a patient with infectious mono.

Occasionally, the EB virus can affect the heart, the brain, the liver, or the respiratory tract. In severe cases, although there's no definite proof that it really works, I prescribe acyclovir (Zovirax), the antiviral agent so useful in herpes (the EB virus is a member of the herpes family). I have a gut feeling it helps. When a mono patient is very sick, steroids may be required.

Chronic Fatigue Syndrome—A Nondisease?

In recent years, a condition called *chronic fatigue syndrome* has received a great deal of attention both in the lay and medical press. Suddenly, as if from nowhere, thousands of individuals, usually between 20 and 40 years of age and previously healthy, began complaining to their doctors of severe, unrelenting fatigue. Some of them were also found to have tender, enlarged glands in the neck, occasional sore throats, and intermittent low grade fever—all persisting for weeks and months. Many of these persons had in common evidence in the blood of Epstein-Barr (EB) virus infection. So this "syndrome" was, at least initially, believed by some doctors to be either a flare-up or a recurrence of infectious mononucleosis—a disease known to be caused by the EB virus. But other doctors disagreed, pointing to the fact that 90 percent of the population has been exposed to EBV at some time or other, and, therefore, would be expected to have increased antibody levels in the blood.

The controversy continues. I've read scientific papers asserting that if one looks carefully enough at individuals with the symptoms of chronic fatigue, other abnormalities to explain their complaints are usually found—anemia, some hidden disease like thyroid malfunction, an insidious infection, a previously undiscovered malignancy, and very frequently some psychiatric disorder, usually depression. But there have been an equal number of reports representing that the chronic fatigue syndrome is indeed a viral infection, not by the EB agent, but by some as yet unidentified virus. That's what I personally believe. Still, whenever a patient complains to me of unexplained chronic fatigue, I undertake a complete workup to exclude all the other possibilities. If I fail to hit pay dirt, I assume I'm dealing with this elusive virus.

There is no specific treatment for the chronic fatigue syndrome, regardless of its cause. In my view, the millions of dollars spent by its desperate victims on special vitamin concoctions administered intravenously are wasted. If you are advised to undergo such therapy—which, incidentally, is usually quite expensive—shout "poppycock" at the top of your lungs and leave the premises! If a thorough checkup reveals no significant abnormalities, you should bask in the reassurance that your symptoms are not life-threatening, and that with rest, a nutritious diet, and time, they will very likely clear up.

INFERTILITY: More Options Than Adoption

Fifteen to 20 percent of couples who want to have children aren't able to—at least not right off. If you and your spouse have worked at it for more than 6 months without success, it's time to consult a fertility specialist. You will *both* undergo a complete evaluation to get to the root of the problem. The potential culprits are many—thyroid malfunction, endometriosis, scarred fallopian tubes, diabetes, abnormality of the sperm, an undescended testicle, varicose veins in the scrotum, hormonal imbalance, pelvic inflammatory disease, a medication you never suspected...for example, in men, the anti-ulcer drug cimetidine (Tagamet), or nitrofurantoin (Furadantin, Macrodantin) used in the treatment of urinary tract infections.

The best therapy for infertility obviously depends on its cause. In some cases it can be cured by something as simple as the man's not wearing tight undershorts, or avoiding saunas and hot baths before sexual intercourse. On the other hand, it may require sophisticated manipulation of the woman's internal hormonal environment or even an operation. But sometimes whatever is rendering either of you infertile may not be correctable. Also, in a substantial number of cases no cause is ever discovered, even after a time-consuming and expensive evaluation. These are depressing and frustrating verdicts to a couple who desperately want a child. If you've received that kind of news, here are some recently available options of which you should be aware.

• *Artificial insemination with your husband's sperm* has a 50 percent chance of success and is your best bet if he is fertile but impotent—that is, he's making plenty of healthy sperm but can't muster the kind of erection needed to "deliver" them. It's also the method of choice if your cervix, for one reason or another, does not allow sperm to gain entry to the uterus.

• *Artificial insemination by a donor's sperm* when your husband's is inadequate also carries with it a 50 percent success rate. Reputable clinics specializing in this field screen the genetics of the contributors very carefully, and also examine the purchased sperm for evidence of infection and AIDS antibodies. *Always have your own doctor check the credentials and reputation of the particular sperm bank you're planning to use;* some of them are not what they're cracked up to be. You may have read about the woman who had her dying husband's sperm stored for use at a later date. After he died, she was inseminated with it and had an uneventful pregnancy. Nine months later she gave birth to a bouncing baby girl. The only problem was that although both parents were white, the baby was black. There must have been some mix-up in the sperm storage procedure because as they say in China, "Two Wongs don't make a white!"

• *Embryo transfer* (ET), also referred to as *in vitro fertilization* (IVF), is a more expensive and complicated way to have your own baby. By contrast to insemination, in which sperm is the weak link in the chain of conception, you will require ET when the fallopian tubes (that lead from the ovary to the uterus) have been so scarred by infection that the egg cannot get through them to get to the uterus. In this technique, one of *your* eggs is removed from *your* ovary with a needle guided by ultrasound. It is then placed in a culture dish where it is fertilized when *your spouse's sperm* is added. The egg is then transferred from the dish and implanted into your own uterus where it continues to grow for the usual 9 months. Embryo transfer has a 20 to 30 percent success rate, and can be attempted as often as necessary. Before your ovary is tapped, however, you should be given clomiphene (Clomid) or follicle-stimulating hormone (Pergonal) to stimulate the development of more eggs to improve the chances of a "hit" by the needle. This technique has now been used for almost 15 years, and has resulted in the birth of more than 25,000 babies in over forty different countries. Leaders of the various faiths—Islamic, Buddhist, Jewish, and Christian—have all approved its use by married couples.

• When the uterus itself is abnormal and cannot nurture a pregnancy, *your fertilized egg* can be implanted into the womb of a "*surrogate mother.*" The genetics are all yours, but the "incubation" is not. The major complication is the possibility that the surrogate will refuse to part with the infant.

• Finally, when you simply aren't making any eggs yourself, a *surrogate mother whose egg is fertilized by your husband's sperm* is yet another option. Unfortunately, this has led to unpleasant legal battles over custody of the child that have considerably dampened many of the pleasures of parenthood so obtained. So get good legal advice before embarking on these last two alternatives.

INFLAMMATORY BOWEL DISEASE: How to Live with Ulcerative Colitis and Crohn's Disease

Don't confuse inflammatory bowel disease (IBD) with the irritable bowel syndrome (IBS—see page 163). There is a world of difference between the *S* and the *D*. Although both disorders leave you feeling miserable, with abdominal pain, gas, diarrhea, constipation, mucus, and the need always to be near a toilet, the *D* is a threat to life—the *S* is not. Irritable bowel syndrome can make you rue the day you were born, but inflammatory bowel disease—which includes two somewhat different disorders, ulcerative colitis and regional ileitis or Crohn's disease—has a serious impact on health, growth, and survival. Crohn's and colitis can be distinguished from each other by the appearance of the affected bowel, which areas are affected, and how deeply the process involves the lining of the colon. It is important that this distinction be made, because outlook and management vary somewhat between these two disorders.

The causes of IBD, which affects some two million persons in the United States alone, remain a mystery. Symptoms usually begin in the early teens, peak between the ages of twelve and forty, and affect both sexes equally. This disease is not nearly as common in Third World nations as it is in "developed" countries, and there is a higher incidence among Jews than among blacks or Asians. If one of your close blood relatives has IBD, chances are one in four that you will (or do), too. I know a family in which every single member except the father has the disease.

The intestinal tract of persons with IBD is *chronically* inflamed. That may be the result of an autoimmune disorder in which the patient's bowel is being "rejected" by the immune system for no good or obvious reason. This rejection process (in which the body views its own bowel with hostility and tries to get rid of it) presumably causes the inflammatory response of pain, diarrhea, cramps, and fever. When an IBD bowel is looked at with a colonoscope or sigmoidoscope, the physical evidence of this reaction can be seen—ulcers, swelling, and scarring are apparent. By contrast, an IBS bowel looks "normal" when viewed through a scope.

Although medication is the mainstay of therapy for IBD, diet plays an important role, too. Many of my patients are able to identify some food that predictably worsens their symptoms, usually a dairy product containing lactose. That's because so many of them lack the enzyme lactase, which is necessary to digest lactose, a sugar present in milk and milk products. Surprisingly, hot or spicy dishes are not usually provoking factors. Indeed, coffee, fat or fried foods, onions, nuts, seeds, and *insoluble fiber* foods are often worse than chili! So try to correlate any flare-up of your own symptoms with what you've been eating.

Some doctors tend to designate anyone with an intestinal problem, upper or lower, as nervous or high strung. I have not found that usually to be the case in IBD, but psychotherapy can help you cope with this affliction, reduce the intensity of your emotional response to it, and so improve the quality of life.

The standard medication for IBD has long been sulfasalazine (Azulfidine), which often improves symptoms (especially when they're mild) and sometimes even induces remissions. Azulfidine consists of two active components—sulfa and an aspirin-like drug called 5-amino-salicylic acid (5-ASA). It is the latter that actually treats the disease; the sulfa only delivers it to the affected area of bowel. But Azulfidine is not always well tolerated. Twenty to 30 percent of those taking it develop headache, nausea, and loss of appetite and generally feel lousy. In addition, anyone who happens also to be allergic to sulfa may come down with hives or other skin rashes, fever, or arthritis and breathing difficulties. Over the years, researchers have looked for ways to get the 5-ASA, the active component of the drug, into the gut *without* attaching it to sulfa—and they have finally succeeded! There is now an *enteric* coated 5-ASA, olsalazine (Dipentum), which reaches the site of the disease lower down in the bowel without first being dissolved in the stomach. Its side effects are fewer than those resulting from Azulfidine, but sometimes there is a rash, bloody diarrhea, fever, or headache. Do not take olsalazine if you are sensitive to salicylates.

Azulfidine remains the first-line treatment for children and anyone else who can tolerate it, but I prefer olsalazine for those sensitive to sulfa. The usual dosage is 500 mg tablets, twice a day—indefinitely. You've got to keep taking it in order to prevent or delay a flare-up.

When IBD involves the very lowest portion of the colon and the rectum, the 5-ASA can be delivered directly via an enema (Rowasa). We used to rely on cortisone enemas, which I now prescribe only when the Rowasa doesn't help.

Some persons with ulcerative colitis and Crohn's disease may respond to metronidazole (Flagyl). Why this drug helps is not clear. So if there is no response to Azulfidine or olsalazine, try Flagyl, 250 mg, 4 times a day for 2 weeks. Alcohol should be avoided during that time because Flagyl and booze don't get along.

If the IBD has not responded to the drugs mentioned above, and the patient feels sick, the next step is steroids taken by mouth (Prednisone). Steroid therapy should be tapered slowly and stopped after it has produced a remission, at which time the Azulfidine or olsalazine is resumed. Taken over the long term, steroids can induce high blood pressure, fluid retention, peptic ulcer, and osteoporosis.

Most patients respond to intensive steroid treatment in about 2 weeks. If you don't, or the symptoms flare up after the hormone has been stopped, there is yet another alternative, a drug called mercaptopurine (Purinethol), which works by blocking the body's inappropriate rejection response to its own bowel. The starting oral dose of 50 mg a day may be increased if necessary, but always *under very close supervision* by a doctor experienced in its use. The main risk of mercaptopurine is damage to the bone marrow.

If IBD fails to respond to medication, surgery, in which the diseased portion of bowel is removed, may become necessary. But before you agree to an operation, make sure that the doctor who has recommended it is a gastroenterologist specializing in IBD. Even if he is, I advise a second opinion because this is major surgery. If there is consensus among your doctors that an operation is your only option, so be it. If you end up with a "bag" into which you must now evacuate because the remaining portion of your bowel is too short to reach the rectum, don't let the prospect of this artificial opening in your abdomen get you into a funk. The newest containers are convenient, odor-free, flat, and, once you get over the initial shock, not at all hard to adjust to.

Because IBD is so common and important a cause of serious disability, it has been the target of a great deal of research. In recent years, several new and better medications have become available for its treatment, and more are on the way. If you have active IBD that has not responded well to whatever treatment you're currently receiving, register with the National Foundation for Ileitis and Colitis, at 444 Park Avenue South, New York, N.Y. 10016, to receive their regular newsletter.

INSECT BITES AND OTHER STINGS:
Please—Bug Off!

Whether or not you're going to get stung or bitten by the mosquito that's dive-bombing you, or the wasp that's eyeing you from inside the windshield of your car as you're driving at 55 miles an hour, is usually, but not always, a crap shoot. For example, any mosquito within striking distance of my wife can and will find, land on, and then bite the tiniest exposed area of her body. I, however, am immune. When we're together, I function as the mosquito's launching pad in its attack on my spouse. No matter where I've been in the world, I have never been bitten. I wear no repellent; I don't even bother to shoo mosquitoes away. They simply have no interest in me.

Now, as far as hornets, wasps, and bees are concerned, that's a whole other story! They chase *me,* but totally ignore *her.* I remember one hornet that literally pursued me for 20 minutes, forcing me indoors. And what an attention span it had! I could see it hanging around on the patio just waiting for me to come out and play. I don't dare wear any aftershave lotion when sitting at a pool; that attracts them, too. Even when I walk barefoot on the grass, something I love to do, I invariably step on one, only to be rewarded by a painful sting.

For a few people, insect bites can be life-threatening. For most of us, however, they are merely an unpleasant experience. The following advice won't lessen your chances of being bitten, but it may save your life if you are.

Some individuals have a hypersensitivity to the venom of certain insects, especially hornets, wasps, and bees. In addition to a very painful reaction at the site of the bite itself, hypersensitive persons often develop palpitations, flushing, sneezing, and shortness of breath a few minutes later. *This is a life-threatening situation* that can lead to collapse and shock. If you are hypersensitive to the bite of these insects, and will be spending time in a rural area, arrange to be desensitized first. That's not a one-shot deal. It means being injected repeatedly with venom until your body builds up an immunity to it. Truly effective desensitization that will protect you against a fatal or near-fatal reaction to a sting requires 5 years of treatment! That may seem like overkill just to be able to go to one picnic, but if you're truly vulnerable, and are stung, that picnic can be your last! *If you have not been desensitized, always carry Ana-Kit with you.* This is a self-injection device that requires a prescription and consists of a syringe preloaded with just the right dose of adrenaline. It can save your life. Make sure you know how to use it, and practice giving yourself "dummy" shots *before* you're bitten. If you're stung, don't wait to see what happens. Use your Ana-Kit, then get to the nearest hospital emergency room as quickly as possible. If there is no Adrenalin available, take any antihistamine that is.

So much for the small percentage of the population that is really at grave risk for these bites. If you're not, and have been attacked by ants, hornets, mosquitoes, wasps, or bees, you're in no danger from a single bite, and obviously don't need to be desensitized or carry Ana-Kit or go to an emergency room if you've been stung.

You may have read about the invasion of the United States by the African bees, which have been traveling north from Mexico at a speed of about 200 miles a year. They've already arrived in Texas, and are expected to replace most of the native bees in the southern, warmer areas of the United States in the next few years. Despite all the fanfare and fear concerning these "killer" bees, they actually contain less venom *per bee* than their local brethren. But they are more aggressive and tend to attack in large numbers. There have been reports of individuals suffering hundreds of bites! While hypersensitive persons are especially vulnerable to the African bee, multiple stings are a threat to everyone. Should you sustain several bites from *any* insect, regardless of your "allergy" profile, see a doctor immediately.

Here's what to do if you are stung by a bee, hornet, or a wasp. First, carefully remove the stinger, making certain not to leave any portion of it behind. To reduce pain, apply an ice cube directly on the spot where you've been bitten, then cover the area with a topical steroid like Cortaid and take any antihistamine pill.

Ticks

There is now a major interest in repelling the tick that carries Lyme disease, an important cause of disability and death in many parts of the world (see page 177). In my opinion, the best overall topical repellent against ticks, mosquitoes, and biting flies is N-diethyl-M-toluamide—better known, thank goodness, as Deet. Apply it to every exposed area of your body and put more on if you are caught in the rain, sweat too much, or have been swimming. I prefer this particular product because it's not as easily wiped off and is less likely to evaporate than are most of its competitors. But as with any medication, there is a risk of allergic or toxic response to Deet, especially since a small amount of it is actually absorbed through the skin. Also, be sure to keep the container away from children, because Deet is extremely toxic when inadvertently swallowed.

Remember that ticks and other insects don't just attack the skin; they also cling to your clothes and can get at you later when it suits them. So if you're in an area heavily populated by insects of any kind, apply a repellent to your clothes as well as to your skin. The one I like best is Permanone tick repellent; it won't stain your clothes or make you smell bad. The combination of Deet on your skin and Permanone on your clothes should virtually eliminate any chance of your being bitten. But, you should still cover your body as completely as you can if you live in a tick-infested area, and are in the habit of walking through the woods or in deep grass.

If you find a tick on your body, remove all of it as soon as possible. If you can, first douse it with rubbing alcohol or turpentine. Grasp it with a pair of tweezers at your skin line, and either pull straight up very gently or twist it out slowly. Make sure not to squeeze the tick's body. That's where the Lyme organism resides, and if you spill its stomach contents onto your skin, you may liberate the bug that can then infect you.

Exotic Bites

• *Spiders* aren't really worth worrying about because although most of them are venomous, the fangs of all but a very few are too short to penetrate the skin. The *brown recluse* and the *black widow* spiders, however, can and do. A *brown recluse* spider bite can be serious. If you're *sure* that's what bit you (the tiny puncture site develops a gray-blue halo around it a few hours after the sting), the best treatment is a steroid injection as soon as possible. Children often react very badly to the sting of the *black widow,* which sometimes causes a numbing pain at the site of the bite, but most adults do not. Treatment is best obtained in a hospital emergency room and should consist either of intravenous calcium, antivenom, or both. Most spider bites in the United States (400 per year) have been reported from Southern California.

• *Scorpion* stings, which you are most likely to get in Arizona or Mexico, are usually harmless. They are best treated, where possible, with the immediate application of a tight dressing or a tourniquet to prevent the toxin from spreading through the body. Although this complication is uncommon, if you're bitten, call the Arizona Poison Control System at 602-626-6018—wherever you happen to be.

- If you've been bitten by a *stingray* while swimming in the ocean, get out of the water and onto the beach. Then wash out the penetrated area with lots of salt water. As soon as you can do so, immerse the limb or whatever area was stung into very hot salt water for about 2 hours. Now look very carefully at the affected area. If you see the fish stinger, *have a doctor remove it.* If left alone, the venomous sting can cause severe pain, nausea, vomiting, and even collapse.
- A *sea urchin* sting is by no means unusual on ocean vacations, virtually anywhere in the world. Here's what to do if you're "elected." Apply some Sea Balm, which you can buy over the counter, to the area, remove the spine, then apply vinegar compresses for at least 2 hours. If not removed, the spine can penetrate into deeper tissues, causing infections.
- If you've been swimming in the ocean and been stung by a *red jellyfish,* flood the site of the bite and the protruding stinger in table vinegar to inactivate the toxin, then scrape the skin free of the stinger. Sea Balm, which consists of a pain killer, topical steroid, and antihistamine, soothes the discomfort.

The pleasures of communing with nature are sometimes marred by all kinds of crawlers and fliers with whom we share the environment. They are generally more tolerant than we humans, but when they're not, they can make life miserable and even threaten it. The best approach is to avoid them where possible, frustrate them by keeping covered if you're vulnerable, carry adrenaline with you if you're hypersensitive—and hope for the best.

INSOMNIA: Your Passport to the Land of Nod

Insomnia is a symptom, not a disease. No one is born destined to become an insomniac. Like any other symptom (pain, itch, cough, fever), there's always a reason for it, the most common being something psychological—you're either worried, anxious, fearful, depressed, or excited. Or, you may be perfectly content, but your *sleep environment* is at fault—your bedroom is too hot, too cold, or poorly ventilated; your bed is too short; the mattress is bad for your back (remember the princess and the pea?). Or, you may have developed bad *bedtime habits,* like trying to catch up on your unfinished office correspondence before turning out the lights; working out on your rowing machine or treadmill (exercise stimulates some people, relaxes others); smoking that last cigarette of the day (the nicotine will keep you awake); even your sex habits can cause insomnia—it soothes most of us, but overstimulates others. Delaying your liaisons until morning may cure your insomnia!

Then there is the question of *medication.* You may be taking something you never thought interfered with sleep—more thyroid supplements than you need, cold remedies containing decongestants like phenylpropanolamine (PPA) or pseudoephedrine, appetite suppressants, alpha methyldopa (Aldomet—a drug used to lower blood pressure); phenytoin (Dilantin) for the prevention of seizures; propranolol (Inderal) or other beta-blockers (for management of various heart problems and hypertension); caffeine or lots of chocolate taken late in the day or during the evening; a diuretic at bedtime (it keeps you hopping to the john all night); wine or brandy at bedtime (it may help you get to sleep, but can also awaken you in the middle of the night—a kind of withdrawal effect). If you've become habituated to sleeping pills and stop them abruptly, you will suffer severe insomnia for several days.

Finally, symptoms due to an *underlying physical condition* that makes you uncomfortable at night will obviously interfere with sleep—shortness of breath due to heart failure; hiatus hernia with acid pouring into your chest from your stomach; the respiratory distress of emphysema or asthma; a chronic cough; a large prostate that sends you to the bathroom every hour; the pain of a cancer or arthritis.

In order to ensure a predictable and restful sleep, you must first control or eliminate any of the above factors that are operative in your case. If that doesn't make a difference, then there are other positive steps you can take. Certain foods help trigger sleep, especially milk, tuna fish, soybeans, turkey, even eggs (try two, especially if your cholesterol is nice and low). These are all rich in *natural L-tryptophan,* a sleep-promoting amino acid that acts on the brain to cause drowsiness, and does so more quickly when combined with a carbohydrate like orange juice. My grandmother knew all about it when she advised us to drink a glass of warm milk at bedtime (her generation wasn't worried about cholesterol) in order to get a good night's sleep. *L-tryptophan supplements* are currently banned in the United States because they have been shown to cause serious health problems. I suspect it will turn out to have been due to some contaminant in the manufacturing process, and once that's straightened out, L-tryptophan will again be available. Until then, don't use it, even if you still have some lying around from before the recall.

Here are a few simple things to try. Take three deep breaths very slowly, exhaling fully each time. After the third exhalation, stop breathing for as long as you comfortably can. Repeat the cycle five or six times. This breathing exercise causes a tranquilizing effect because of the accumulation of carbon dioxide in the blood. A warm bath may help, too. Also, look into some of the sleeping aids like self-hypnosis recordings. If you live near an airport or highway, try a sound masking device that can replace disturbing outside noises with the soothing sound of surf.

The point of all the above is that *you should exhaust every possible option before resorting to pills.* But sometimes, personal crises beyond your control make sleep well-nigh impossible, at least temporarily, and you really need to take something. Stay with an over-the-counter antihistamine first—one that does not require a prescription. The FDA has approved three products for this purpose—Sominex, Nytol, and Compōz. *But remember, every drug, even one you can buy freely and without restriction, has potential side effects.* Antihistamines can cause a rash, dry mouth, interfere with urination in men with enlarged prostates, and occasionally leave you confused and delirious. So use them only for the shortest time necessary and in the doses recommended on the bottle.

If at this point you *still* can't sleep, I recommend an antidepressant like amitriptyline (*Elavil*), 25 to 100 mg at bedtime, *when the insomnia is due to depression.* On the other hand, if you are anxious, tense, nervous, but *not depressed,* I have found the *benzodiazepines* to be the safest and most effective drugs for the purpose. I prefer diazepam (Valium), 5 mg, flurazepam (Dalmane), 15 mg, or triazolam (Halcion), 0.25 mg. However, none of these are free of possible toxicity. For example, Dalmane may leave you with a hangover, while Halcion, which some travelers use on long overnight flights, has been known to cause amnesia for events that occur *after* you've taken it (called anterograde amnesia). Regardless of which benzodiazepine you use, do not continue it for longer than 4 weeks. Take it with meals, but *never with alcohol;* the combination can damage the brain. Although the benzodiazepines are *relatively* safe, they can cause dependency. Even so, I much prefer them to the barbiturates—phenobarbital, pentobarbital (Nembutal), and secobarbital (Seconal), all of which used to be so popular years ago. *Avoid all sedatives, including the benzodiazepines, during the first 3 months of pregnancy* and while you're nursing because these drugs are excreted in the milk.

You may safely stop any benzodiazepine abruptly if you've been taking it for less than 2 weeks. Beyond that, it's best to taper the drug because sudden cessation can cause withdrawal symptoms.

A final word. No matter how "great" you think a sleeping pill is, remember that *drug-induced sleep is not natural.* It reduces dream activity, and a healthy psyche needs dreams. So here's the bottom line. Don't depend on *any* medication for longer than you absolutely must, and if at all possible, don't use it for more than a week or two. If you have trouble sleeping without one of the "aids," then seriously consider discussing the matter with a trained therapist.

IRRITABLE BOWEL SYNDROME: When You're Sick, But All the Tests Are Normal!

One of my patients, a woman in her early fifties, has had recurrent diarrhea for as long as she can remember. When she gets the urge to go, she has to move fast. After she is through, nature often calls again in an hour or two. Sometimes it turns out to be only gas, but she never knows for sure—beforehand. The diarrhea may stop for 3 or 4 days, only to be followed by constipation, so that when she does finally "go" she eliminates only a few tiny hard pellets. She almost always has a crampy pain in the left lower part of her belly.

This woman knows more doctors than do most malpractice lawyers! You name it, she has seen them—internists, gastroenterologists, psychiatrists, parasitologists and infectious disease specialists, neurologists, chiropractors, osteopaths, allergists, and a host of holistic types. She has been examined from head to toe, her stool has been repeatedly cultured for every parasite and bacteria known to man, and looked at under the microscope; she has undergone countless biopsies, barium enemas, skin tests, colonoscopies, CT scans, and psychological evaluations—all of which were invariably reported as "normal." Every specialist has given her the same "good news," that she has "nothing more than an irritable bowel" to which she simply must "adjust."

This woman's symptoms are shared by hundreds of thousands of individuals. The irritable bowel syndrome (IBS), also known as nervous bowel or spastic colon, is twice as common in females as in males. It may begin as early as the teens, or not before midlife. The clinical picture may vary—some days or weeks are better than others, but the individual with IBS never really feels good for any length of time. Most are high-strung, tense, and nervous, but who wouldn't be with this constellation of symptoms? I happen to believe that, despite all the "negative" tests, the irritable bowel is a real disorder whose cause we simply do not *yet* understand. Those affected do not lose weight, they're rarely anemic, they don't often have blood in their stool, and they never have fever—but they are sick and miserable nevertheless!

At the moment, there is no "best" treatment for IBS, but certain steps can make life more tolerable. What is needed first and foremost is the reassurance that IBS, as disabling as it is, is not a threat to life, that it does not lead to Crohn's disease, ulcerative colitis, or cancer. Having established that, the next step is dietary advice and the proper use of antispasm medication.

If you have IBS, *avoid gas-forming foods* like dairy products, cauliflower, beans, bananas, brussels sprouts, onions, cabbage, and even bagels. Eat all the meat, fish, chicken, lettuce, popcorn, nuts, and even potato chips you want. I also recommend a *lactose-free diet* because I believe many IBS sufferers are lactase-deficient. It's important, too, to increase the amount of dietary fiber you eat so that the stools are more bulky, water-logged, and easier to pass. Oat bran, wheat bran, and 1 to 3 tablespoons of Metamucil (psyllium) daily can make a big difference, too. Remember, psyllium is a fiber, and good for you—not a laxative.

If dietary changes don't work, I recommend antispasm drugs. I prefer Bentyl, Librax, and Donnatal, usually taken 3 times a day before meals. Some patients have taken them for years. If diarrhea due to the IBS becomes troublesome, you may need as many as 6 Lomotil or Imodium a day, again, sometimes indefinitely. When any of my patients with IBS become depressed (understandably), I often suggest amytriptyline (Elavil), 25 mg at bedtime for a few days (but not as a steady "diet").

If you have an irritable bowel, remember—for whatever consolation it may be—that there are countless thousands like you who function normally, secure in the knowledge that their disorder carries with it no special risks other than misery and social embarrassment.

ITCHING: More Than Skin Deep

The itch that's driving you crazy can stem from a variety of causes—everything from poison ivy to cancer—including bites, rashes, drugs, allergies, and even psychiatric disorders. It can affect your entire body, or only a small area; although usually external and limited to the skin, it can also be internal (vaginal, rectal), and it may last for hours, days, weeks, or months.

Fever may make you uncomfortable, but it also hurts the organisms that are infecting you; pain causes you grief, but it alerts you to danger. An itch, however, has no redeeming features whatsoever.

The best way to control an itch depends on its cause. That is not to say you must identify *exactly* what nasty little insect it was that bit you—mosquito or bedbug—but you should at least know that your itch is the result of an insect bite and not, say, a food allergy. Here's what to do with that information.

If you're itching from a visible rash or can see bites on the skin, apply something *topically* first. I prefer a wonderfully cooling and soothing calamine preparation that contains either menthol or phenol, something like Caladryl cream or lotion, which you can buy readymade without a prescription. Cold water compresses every few hours will help, too, but they're not nearly as convenient. There are also several effective steroid lotions, creams, and ointments available over the counter if you need them, such as Cortaid. They are not as potent as what your doctor can prescribe, but try them before you go to the "heavy stuff."

If the itch involves too large an area of the body on which to apply a cream, then luxuriate in a cornstarch or Aveeno bath for half an hour, 2 or 3 times a day. If your symptoms don't clear up within a week, then consult your doctor. Do not continue to apply topical steroids indefinitely *even if they work,* because they may leave your skin thin, wrinkled, and discolored. Use them on your face only occasionally and for just a few hours.

One word of caution. Most topical preparations are absorbed through the skin to some extent, so *never* apply them exept in the tiniest amounts to children under two years of age. Their skin is very thin to begin with and more easily penetrated. If you need something topical for your infant, use *benzocaine* in a 5 percent cream. It's an effective local anesthetic that does not require a prescription and has been used safely for the relief of itching for many, many years—in the young and the old alike.

An *itch* can also be due to some *internal* problem (thyroid trouble, obstruction of the bile ducts, or certain types of cancer), in which case it is apt to be widespread, with a perfectly normal-looking skin except for your scratch marks. Here, topical preparations are not much help, but oral antihistamines frequently are. They should be your first choice while you are being worked up to determine the underlying cause. I have found Atarax (hydroxyzine), 10 mg, 3 times a day, the best of the lot, but Seldane (terfenadine) is less apt to leave you feeling drowsy and so should be tried during the day. (Remember, however, if you have a big prostate that's getting you up two or three times a night to empty your bladder, *any* antihistamine can make matters worse.) Older patients should also take 50 mg of Sinequan (doxepin) at bedtime if the itch interferes with sleep. All of these agents can be used until the reason for the itch is established and removed.

If you're an allergic type and certain foods cause hives that itch like the devil, try cimetidine (Tagamet). It's extremely effective, especially when given by injection during the acute phase. If that doesn't work, you can always fall back on the antihistamines.

The most common causes of *vaginal itching* are local infections (like trichomoniasis), yeast overgrowth (from having taken antibiotics, or because you're diabetic), or a dry vagina (in menopausal women who lack estrogens). A povidone iodine douche (Betadine) will afford temporary relief until you get to the root of the problem. If you turn out to have trichomoniasis, Flagyl (metronidazole), 250 mg by mouth, 4 times a day for 7 days, will cure you. *Do not consume any alcohol in any form* while you're on this drug; the interaction can be horrendous! Make sure, too, that your spouse or boyfriend is also treated with Flagyl at the same time, because he's probably harboring the organism as well and will continue to reinfect you unless he's also cured.

Suspect *yeast* as the cause of your vaginal itch if you're pregnant, are on the Pill, suffer from some immune disorder, have been taking antibiotics, or are diabetic. Antibiotics interfere with the "normal flora" in the vagina—the bacteria that keep fungus at bay. Any antibiotic, even when used for only a few days, may permit fungus to grow unchecked, leaving you with an itch. Such infections are best treated with a specific antifungal medication like nystatin (Mycostatin) in vaginal suppository form. An itch due to a dry vagina in the *menopausal* female responds well to the application of an estrogen cream such as Premarin (conjugated estrogens). A diabetic female spilling lots of sugar in the urine is also vulnerable to vaginal itching, and in such cases it is important to lower the blood sugar level by better diabetic control.

Rectal itching is almost always caused by internal hemorrhoids. The best treatment in the long run is to get rid of them (see page 125). But until you do, you should avoid foods that irritate the rectal area, like coffee, spices, and tomato juice, and use a cream or suppository (Anusol-HC) that contains a local anesthetic and some hydrocortisone. If the itch is so intense that you can't keep your fingers away, protect the affected area from your fingernails by wearing light cotton gloves when you go to bed.

Whenever you have a persistent itch, whether it's a contact dermatitis from a new soap or perfume or a different fabric you're wearing, or a reaction to some shellfish you've eaten, or an insect bite, have your doctor make the precise diagnosis and prescribe specific treatment. But in the meantime, the topical and other preparations recommended above will usually tide you over.

KIDNEY STONES: More Shocking Than Surgery

If you suddenly develop waves of excruciating pain that start in the small of your back, move around to the front of the abdomen, and shoot into the groin, you probably have a kidney stone. That's even more likely if you're a man, since the incidence of stones is three times greater in men than it is in women. The presence of blood in the urine—either visible to the naked eye, or more usually under the microscope—and seeing the stone on an X ray or a sonogram clinches the diagnosis.

At least one person among every ten reading these pages has already had, or will in the future have, a kidney stone. Of the one million or so painful attacks suffered each year in this country alone, 300,000 are severe enough to require admission to a hospital. Ask anybody who has had one, and he'll tell you that the best treatment is that which provides immediate relief from pain—usually an injection of a narcotic like morphine or meperidine (Demerol). This is no time for a couple of aspirin or Tylenol!

Once you're in less agony, the next objective is to rid you of the stone itself. The best way to do that depends on its size and location in the urinary system. A stone may be located within the kidney itself, or anywhere in the ureter, the duct that carries the urine from the kidney to the bladder. The smaller the stone, and the further down it is situated, the greater the likelihood that it will exit your body on its own. It's best to wait for that to happen. But if the stone is stuck high up, or the pain remains unbearable, or there is evidence of urinary bleeding or infection, it should be removed. Not so long ago the only way to do that was by surgery. Today, however, 90 percent of all kidney stones can be successfully eliminated without an operation by means of *extracorporeal shock wave lithotripsy (ESWL)*. This is how it's done. You sit in a water bath in which a shock wave is produced by the discharge of a high-voltage, rapid-fire spark plug. The energy generated by this shock is directed to the stone by a brass reflector, and its force shatters the stone on impact. You then expel its tiny fragments in the urine during the next few hours or days. ESWL is great, but not if you are pregnant, have a blood-clotting problem, an aneurysm of the aorta, or there is lots of calcium in the arteries near the kidney. In such circumstances, the shock can shatter not only the stone but may injure these other neighboring structures as well. Another situation that makes lithotripsy difficult, sometimes even impossible, is obesity (if you weigh 300 pounds or more), because the shock wave has too much fat to traverse before it can effectively hit the stone.

It is not usually necessary for you to be admitted to the hospital to receive ESWL. It can be done in an out-patient clinic in the morning and you can leave the same night, or at the latest, the following day. After the procedure, you should drink at least 2 quarts of water a day until your urinary symptoms have cleared up and all the stone fragments have been passed. Filter your urine so that the gravel it contains can be analyzed to determine the composition of your stone—information that permits measures to reduce the likelihood of recurrence.

Even though ESWL is safer and certainly easier on you and your pocketbook than an operation, there is some suspicion that it may lead to high blood pressure later on. So once you've had lithotripsy, your blood pressure should be checked at least two or three times a year thereafter.

If you are not a suitable candidate for ESWL for any reason, your kidney stone can be fragmented by an *ultrasonic or electrical technique* instead. Here's how that's done. A narrow, hollow probe is inserted into the urethra and up the urinary tract, or through a small incision in the side of the body, to make actual physical contact with the stone. A burst of ultrasound energy is then generated that shatters the stone. Laser energy can also be used to achieve such fragmentation, but in both methods, direct contact of the instrument with the stone itself is necessary.

Stones within the kidney can also be extracted by means of an instrument passed through the skin directly into the kidney, where it is wrapped around the stone like a lasso. This procedure requires anesthesia. If the stone is down far enough in the kidney duct (ureter), it can be snared from below, through the urethra, with a "basket"-shaped device.

If you have a kidney stone, you should be checked out for the major conditions that can cause it. These include a tumor of the parathyroid glands (parathyroid adenoma), too much vitamin D, gout, and leukemia. The *parathyroid adenoma* is a benign tumor that produces too much parathyroid hormone, which sucks calcium out of the bones and into the urine where it forms stones. It is easily diagnosed by means of a simple blood test, which should be done on everyone with kidney stones made of calcium. *Excessive vitamin D* increases the amount of calcium absorbed by the gut from the food you eat, which then ends up in the kidney as stones. In persons with *gout*, the stones are composed of *uric acid,* not calcium. Treatment consists of medication to lower the blood level of uric acid and lots of fluid to help the stones to dissolve.

If you've had an attack of kidney stones, don't just sit back and hope that it won't happen again. Chances are greater than one in two that you will suffer another one within 5 to 10 years. You can reduce that risk by following these guidelines. Get into the habit of *drinking lots of water*—for the rest of your life. This is especially important if you live in a hot climate. When you perspire, and lose large amounts of fluid that are not replaced, the urine becomes more concentrated. The minerals it contains, which normally remain dissolved, then tend to solidify into everything from gravel to full-blown stones. So replacement of fluid loss is step number one. But you must also *avoid foods rich in the substances of which your stone was made.* For example, if it contained mostly *calcium,* then reduce your intake of dairy products and avoid calcium-rich antacids and supplements. If your stone was predominantly *oxalates,* keep away from too many green vegetables and meats, and avoid chocolate and iced tea. If *uric acid* was the main component, cut down on grapes, instant coffee, berries, citrus fruits, juices, vegetables, and caviar (that shouldn't be too hard to do at the current price of $1,000 for a 14-ounce tin!).

In addition to these dietary measures, there are other steps and medication that can help prevent recurrences. If your calcium stones are due to a parathyroid adenoma, the best treatment is obviously to remove the hormone-producing tumor. Calcium stones due to other causes may be prevented by a diuretic such as hydrochlorothiazide (HydroDIURIL), 25 mg a day.

Allopurinol (Zyloprim), 300 mg a day, will lower a high uric acid level, but make sure to take this drug with plenty of water. Potassium citrate, 60 to 80 milliequivalents (you can figure this out quite easily from the strength listed on the bottle) a day, will render the urine less acid, and should also be taken when you have recurrent uric acid stones.

In summary, remember that most kidney stones pass spontaneously. Those that don't rarely require an operation anymore. The most popular nonoperative way to remove them is by shock waves (lithotripsy). Long-term dietary and medical management can help prevent recurrences. Kidney stones usually reflect an underlying disorder that should be identified and corrected.

LARYNGITIS: Never Whisper

The ability to make sounds, to speak or sing, depends on the normal function of the vocal cords, a curtain-like structure that opens and closes by various degrees depending on the noise you want to produce. When any part of this apparatus is paralyzed and cannot move, becomes swollen, thickened, or develops growths of one kind or another, you "don't sound like yourself."

Losing your voice or being hoarse for a few days is not one of life's major tragedies. It often happens after a cold, or prolonged exuberant support of the home team, or if you're a chronic "voice abuser" (preacher, politician, auctioneer, or frenetic sportscaster). But, the *sudden* onset of hoarseness (laryngitis) for no apparent reason may occasionally reflect a life-threatening situation. For example, should your previously healthy child go out to play and come back hoarse and wheezing, he or she may have inhaled some object that has become stuck between the vocal cords. That's a prime emergency requiring immediate medical attention. But, don't confuse this with the wheezing that some asthmatic children develop after exercise or exposure to cold. The tipoff to the presence of an obstruction that must be removed is the hoarseness. Asthmatics wheeze, but their voice is intact.

You may also become hoarse:

- If you've been intubated during an operation—that is, a tube has been put down your throat during anesthesia in order to get oxygen into your lungs, and it has inadvertently damaged the vocal cords.

- If your vocal cords have become thickened, harbor a cancer, or sprouted a polyp.
- If the nerves that open and close the vocal cords have been paralyzed by pressure from a growth or other abnormal structure (like an aneurysm) in your chest.

Hoarseness for longer than two weeks demands an examination by a doctor. Here is the best way to treat what he finds:

- When you have *acute laryngitis* due to a cold or some other viral infection, your speech will usually return to normal in seven to ten days without medication, but it sometimes takes longer. You can speed things up by using your voice sparingly. Don't talk unless you absolutely must. And here is something that even some doctors don't know: *Do not whisper.* Far from resting the vocal cords, whispering actually strains them more than speaking does. Nor should you waste your time gargling. The solution bubbling up in your throat may make impressive sounds, but does nothing for the vocal cords!
- Surround yourself with *cold-mist humidifiers* or vaporizers whenever and wherever you can. That may not be possible at work but certainly should be at home, especially in the bedroom. There's nothing harder on the vocal cords than an arid environment. By the same token, avoid decongestants and antihistamines, both of which worsen the laryngitis by drying out the vocal cord tissues.
- If you're hoarse and also have a lingering dry cough after a cold, suppress the cough with dextromethorphan or codeine (see page 82). Hacking away leaves the vocal cords swollen and perpetuates the hoarseness.
- If you're hoarse after a cold or from vocal abuse, but don't have fever, antibiotics are a waste of time and money.
- If you've been hoarse for more than two weeks as a result of vocal abuse, heavy cigarette smoking, or exposure to other irritants or toxic fumes, your cords may have become thickened or covered with nodules or polyps. Don't rush to have them surgically removed. Instead, humidify your environment; then enroll in a *voice therapy program,* which is available in many cities and medical centers throughout the country. (If you live in the New York City area, look into the facility at the Lenox Hill Hospital—a particularly good one.) Experts in vocal dynamics can teach you how to use your voice properly. If you allow your vocal cords to be surgically stripped, or the nodules or polyps removed (by operation or

laser) without first correcting how you speak, you not only run the risk of postoperative scarring, but the problem is very likely to recur in a matter of months.

- If you are a cigarette smoker and have chronic laryngitis, you are at high risk for cancer of the larynx, which occurs almost exclusively in smokers. Such a cancer can be cured, if detected in time, by either radiation or surgery, or a combination of both.
- If your vocal cords were damaged or paralyzed by prolonged intubation, or by a direct blow to your neck, or after an operation near the vocal cords (on the thyroid or parathyroid glands in the neck), newer surgical techniques can often restore near-normal vocal function. However, don't expect to sing like Pavarotti or Sutherland, except if that's what you sounded like before it all began!
- The change in the timbre of your voice may be the result of a medical illness that does not originate in the vocal cords. For example, if you have very low thyroid function, the cords swell so that the pitch of the voice is considerably lower than normal. Replacing the missing thyroid hormone will have you singing soprano again in no time—in the shower if not at La Scala or the Met.

In summary, here are the salient facts to remember about hoarseness. The most common cause is a viral infection like the common cold in which case the hoarseness will disappear on its own. You can speed the process by resting your voice and not whispering. Keep away from decongestants because they dry the vocal cord tissues and prolong the irritation. If the laryngitis persists beyond a couple of weeks, have a doctor take a direct look at the vocal cords to determine the cause of the problem. There are programs designed to teach you how to speak properly if you have *chronic* laryngitis due to vocal abuse. Thickened cords, polyps, and benign nodules can all be treated surgically or by laser. Cancers can often be cured by either surgery or radiation—or both. Vocal cords accidentally damaged during anesthesia or surgery can now be successfully repaired.

LEG CRAMPS: Doing the Jig at Two A.M.

How often does this happen to you? You go to bed without a complaint or worry in the world, but three hours later you develop painful cramps in one leg. The exquisitely tender spasm may affect the calf or the thigh, or some toes may decide to point north while the others go south. Interestingly, these nocturnal cramps rarely involve the fingers or arms. Your first objective is to deal with the problem without leaving your deliciously warm bed, so you press your foot against the bed frame to line your toes up. That may do the trick, but more often than not, as tired as you are, you simply have to get up out of bed and walk, dance, or exert some other pressure on the painful limb before the spasm subsides.

The reason for these cramps may not be obvious. They usually occur in persons over forty whose circulation is good, and who are taking no medication. But if you're on a diuretic (water pill), the potassium and magnesium levels in your blood may drop and that's enough to do it. Some patients tell me that an air conditioner blowing on their exposed feet induces spasm of the toes. If you've strained your leg muscles during the day or while exercising at bedtime, nocturnal cramps sometimes result, too. Regardless of the cause, here's the best way to treat such leg cramps:

- Before resorting to any drugs, try wearing short socks (booties) to just above your ankles when you go to bed. Keeping the feet warm in this way may reduce the frequency of cramping.
- Since the cramps may be due to an accumulation of fluid in the lower extremities, keep your legs slightly elevated, thereby increasing the return flow of blood in your veins. If that doesn't work, then do just the opposite. Elevate the *head* of the bed!
- If neither of these measures is successful, take a 260 mg tablet of quinine at bedtime. Available without a prescription, this is probably the most effective treatment now available for leg cramps, although no one seems to know how it works! But first make sure you can tolerate quinine. Not everyone can. It is the active ingredient in tonic water. The few case reports of this drug's damaging the bone marrow in some sensitive individuals are no reason for you not to try it.

- If the quinine doesn't work, try diphenhydramine (Benadryl), 25 or 50 mg at bedtime. But take either the quinine *or* the Benadryl, not both. They are no more effective together than either is singly.
- If neither quinine nor Benadryl affords relief, ask your doctor for verapamil (Calan, Isoptin), 120 mg at bedtime, and try it for at least a week. It sometimes works, again, for reasons that remain mysterious.
- If you're taking a diuretic, have your potassium and magnesium blood levels checked. A decrease in either may cause nocturnal leg cramps. Correcting the deficiency will usually alleviate the symptoms.

LUNG CANCER: Timely Treatment May Save Your Life

Cancer anywhere is a serious business, but a malignancy of the lung is one of the worst. Pulmonary cancer accounts for more than 120,000 deaths every year, and is the most common cause of death from malignancy in both sexes. The cure rate is low even under the best of circumstances, and the level of suffering is high. But cure *is* possible. What you learn in the following pages can maximize your chances of surviving this dreadful disease.

There are three key determinants of survival from lung cancer:

- The first is to prevent it, and that means *no smoking*—ever—or if you do smoke, *stop now*. It's probably not too late if you haven't yet developed such a cancer.
- Then there is *early detection* by routine chest X rays *before* the onset of symptoms. If the tumor is found soon enough, it can sometimes be surgically removed and a cure achieved, especially among the elderly. I have several patients with such a happy outcome in my own practice, as do many of my colleagues.
- The specific cell type of the cancer will, to a great extent, determine whether the best chance for cure is with radiation, chemotherapy, surgery, or some combination of the three.

Kicking the cigarette habit is entirely up to you. No one can stop smoking in your behalf. It's not easy, especially if you're already hooked on cigarettes, but it *can* be done. It does take a great deal of motivation, willpower, and help (hypnosis, certain medications, nicotine gum, special filters, Smokers' Anonymous, and so on).

With regard to early detection, doctors at the Mayo Clinic, Johns Hopkins, and Memorial Sloan-Kettering Cancer Center began a study in 1971 of 3,000 cigarette-smoking men over the age of forty-five to determine whether, and how often, an annual routine chest X ray would detect lung cancer. On the first screening film, 223 cancers that would otherwise not have been detected were found. Almost half of them were in a stage early enough to be cured surgically. After 5 years, three quarters of this group were still alive. In other words, *82 lives were saved right off the bat at a cost of 30,000 films (that comes to anywhere between $7,500 and $10,000 per life).* Too high a price? I don't happen to think so. So this is what I do in my own practice. Anyone who smokes more than fifteen cigarettes a day, male or female, gets a chest X ray every year, as do those who at some time or other were exposed to asbestos or other cancer-causing substances.

If a screening chest X ray reveals a suspicious "spot" on your lungs, before making your will, have *all* your previous chest X rays in the past 3 or 4 years reviewed very carefully. A tiny "abnormality" often turns out to be nothing more than an old healed infection. If it was present more than 3 years earlier, you can almost certainly forget about it. Cancer of the lung just doesn't sit around waiting to be diagnosed for that length of time.

But if the abnormality is new, you must now determine what it is—cancer, a fungus, TB, a patch of pneumonia, or some other recent infection. In such cases I assume the most optimistic possibility—namely, a treatable cause like pneumonia—and I prescribe an antibiotic, usually erythromycin, for about 2 weeks. I then re-evaluate the patient to see whether the "spot" is still there. You'd be surprised how many of them have disappeared the second time around.

If, however, after a short course of antibiotics the abnormality persists, I arrange for a CT scan of the chest. That often provides the additional information necessary to distinguish malignancy from other causes. If it doesn't, we must obtain a "tissue diagnosis," the ultimate test as to whether or not you have lung cancer, and if so, the specific type it happens to be. A piece of the area in question is taken from your lung and studied under the microscope.

The outlook and treatment of lung cancer depends a great deal on the specific appearance of its cells. For example, the malignancy may be "secondary"—that is, it has arisen elsewhere in the body (in the kidney, breast, bone, prostate, or intestinal tract) and spread to the lungs. The original focus may have been silent and undiagnosed so that the first indication of any trouble is the abnormal chest film. In most cases, the presence of such a metastatic or traveling cancer is grim, but not necessarily hopeless. For example, cancer of the testicle that has reached the lungs can be cured.

The cancer may be *primary*—that is, originating in the lungs. Here again, the microscopic appearance is critical, as is the age of the patient. Lung cancers are divided into two types—small (oat) cell and non-oat cell. The oat cell cancer, which accounts for about 25 percent of primary lung tumors, can be cured by a combination of X-ray treatment and chemotherapy if detected early enough. In my experience, however, virtually every other lung cancer (non-oat cell) can be cured only by surgical removal. The chances of this happening are greater, believe it or not, the *older* you are—presumably because these cancers do not grow or spread as quickly as they do in younger persons.

Here is what I recommend to my own patients with lung cancer. If the cancer, regardless of its cell type, appears *not* to have spread and lung function is good, then surgery *should* be done without delay, and in all age groups, even in the elderly who are otherwise in good health. Oat cell cancers are treated with chemotherapy and radiation, as recommended by an oncologist.

If CT scans and other tests indicate that it's too late for surgery (because the cancer has already spread to the glands in the opposite side of the chest or elsewhere), I leave it alone if it's not causing any symptoms. I prescribe radiation, surgery, or chemotherapy *only* if the *incurable* cancer is causing symptoms. The purpose of intervention at this point is to make the person *feel better*. When a patient with advanced lung cancer has no complaints, I have them do nothing. This approach is not shared by all or even most doctors. The majority of my colleagues go full-blast with radiation, surgery, and toxic chemotherapy in the hope of obtaining a cure, which rarely, if ever, happens at that stage. Such treatment, in my experience, often accelerates or prolongs suffering.

Whether or not to intervene in hopeless cases with treatment that is likely to make one sick, just *because the tumor is there,* is a decision in which the patient and his or her family should participate. Some of the factors to consider before reaching a final conclusion are age (the younger the patient, the more justified one is in an aggressive approach), the presence of other life-threatening diseases, and whether or not lung function was normal before the cancer was diagnosed.

LYME DISEASE: Suddenly—An Epidemic!

Lyme disease has nothing to do with eating the fruit that looks like a lemon but is green. Lyme is the name of the town in Connecticut where in 1975 the association between a tick bite and a very serious disease was first made. It's interesting how it all came about. Some residents in that community were impressed and concerned by the fact that so many of the adults and children living there had "rheumatoid arthritis." A Yale University scientist, Dr. Allen Steere, solved the problem. He found that the condition the Lyme residents had was caused by a *spirochete,* an infectious organism that belongs to the same family of germs that give syphilis. Lyme disease is not limited to the northeastern United States; it occurs throughout this country, and abroad as well. In fact, the typical symptoms of this infection were actually described as early as 1887 in Australia, but no one knew what caused them.

Lyme is tricky to diagnose unless you have a high index of suspicion for it. Thanks to a heightened public awareness, most diagnoses are made, or at least suggested, by patients themselves. (Oh, the power of the media!)

Here's how you get Lyme disease. First, the spirochete infects a warm-blooded mammal. In the woods, it's usually a white-tailed deer or a white-footed mouse, but it can also be your dog or any other pet that has the run of your grounds, especially between May and August. The spirochete frolics happily in its host's bloodstream until a tiny, hungry tick bites the animal and swallows blood containing the organism. The tick, now infected, lies in wait for *you* as you romp through the grass or the woods with lots of bare skin showing. It lands on you, bites you (of which fact you may be unaware), and transmits to you the spirochete that it acquired when it bit the host animal. So, the spirochete is the villain, your dog or a deer is the host, the tick is the vector or carrier, and you are the fall guy.

How do you know you have Lyme disease? In the classical sequence of events, you find a tiny tick on your body. If you do, assume you have been bitten. Within 2 days, but as late as 3 weeks, if the tick itself was infected (and not every tick is), you may develop a small red pimple that later spreads into a characteristic ringlike or bull's eye rash lasting for about 2 to 4 weeks. You may also become unusually tired during that time, have some fever, aches and pains, chills, headaches, and a stiff neck, all of which ultimately disappear without a trace. So far, your Lyme disease is still in the early stage, and that's the best time to treat it. If it's not, weeks or months later the second phase of the disease sets in with a variety of different symptoms. You may have *neurological symptoms,* in which any group of muscles can become paralyzed, most often the facial muscles so that you think you've got Bell's Palsy; or *cardiac symptoms,* with palpitations reflecting a rhythm disturbance; or *arthritis-like symptoms,* with generalized pain in the joints, muscles, or bones. Sometimes, symptoms do not appear for months or years. In this late stage, because the possibility of Lyme disease is long forgotten, you may be wrongly diagnosed as having rheumatoid arthritis, multiple sclerosis, or even Alzheimer's disease.

If it's all so straightforward, why the diagnostic problems? There aren't any *if* you've seen the tick or recognize the rash and then develop heart trouble, nervous problems, or arthritis and make the association. But here's the rub. The deer tick is so tiny that you may not ever notice it. (The dog tick, on the other hand, may be the size of a pea.) To make matters even more difficult, the telltale rash does not always appear. So now, seemingly out of the blue, you simply come down with a bad case of arthritis, a heart rhythm problem, or a neurological disorder, all of which are very common anyway. In these circumstances, few doctors and fewer patients think of Lyme disease. Failure to institute proper treatment can be crippling or, rarely, even fatal down the line! So remember, if you suddenly become sick with the above symptoms, insist on a test for Lyme disease no matter where you live and whether or not you ever picked a tick off your body. Make sure, however, that the blood is sent to a laboratory experienced in doing such analyses. It's not an easy examination to perform. If you really want to impress your doctor, insist that the lab use the *IgM-capture enzyme immunoassay,* or better still, the *ELISA analysis.* That's the most sensitive test for Lyme disease. But no matter which procedure is chosen, a negative result does not necessarily exclude the possibility of Lyme because it takes weeks for the antibodies one looks for in the test to develop.

Once the diagnosis is made (hopefully within 30 days of your having been infected), tetracycline, 500 mg, 4 times a day for 20 days, will almost always eradicate the disease. If you're pregnant, breast feeding, or your infected child is younger than eight years or still has his or her baby (deciduous) teeth, tetracycline will stain them, so you're better off with penicillin (Pen V), 250 mg to 500 mg by mouth, 4 times a day for 20 days, or Amoxicillin, 500 mg, 3 times a day. If you're allergic to penicillin and also can't take tetracycline, erythromycin—333 mg of E-Mycin or PCE, in the same dosage 3 times a day—is also effective. Late complications of Lyme disease involving the heart, nervous system, or joints require either intravenous penicillin or ceftriaxone.

Supposing you find a deer tick on your body, should you be treated for Lyme? Since it takes two to three weeks for the Lyme disease test to become positive, should you wait for the results, or is it more prudent to start the tetracycline immediately? There's no agreement among specialists as to what course to follow. My own recommendation is to go ahead with the treatment even though the infection itself has not been confirmed. Other doctors will tell you to make note of the fact that you were bitten, and remain on the alert for the next few weeks. Then, at the first suspicious symptom, get yourself tested and begin treatment. That's not a bad approach either.

Be especially alert to the possibility of Lyme disease if you are pregnant, because if you are infected and are not treated, your infant may be born with various birth defects.

The consequences of Lyme disease can be serious and permanent if treatment is not begun as early as possible. That treatment, as you have seen, is very straightforward; the key is early diagnosis and treatment, which requires a high index of suspicion for the disease followed by its confirmation and eradication. Also be aware that you can get it more than once. Unlike viral disease, Lyme does not leave you immune to future infection.

MALARIA: This Section May Be More Important to You Than You Think

Malaria is the most common infection in the world today. Would you believe 200 to 300 million cases every year, and 2 to 3 million deaths? It was referred to in the Bible as "the burning ague," and has been known to man throughout the ages, even affecting several American presidents—Washington, Jefferson, Madison, Harrison, Taylor, Pierce, and Lincoln.

Even before its cause was understood, malaria was known to be prevalent in swampy tropical marshlands, and was given its name by an Italian scientist who believed it to be caused by "bad air". It wasn't until this century that it was found to be due to a parasite harbored by the female mosquito, which infects humans with its bite.

Malaria is a potential threat, and of practical importance to everyone everywhere, even those who reside in a temperate zone, but especially to persons living in Africa, Asia, and South America. So if you're planning to go to any of those countries, there are drugs that you *must* take *before* you leave home, while you're there, and *after* you get back.

Years ago, the best way to prevent malaria was to take one tablet (500 mg) of chloroquine (Aralen) a week, starting 14 days before you embarked on your trip, continued weekly while you were away, and every 7 days for 6 weeks after you came back. But in recent years malaria parasites have become *chloroquine-resistant* in so many areas of the world that you're now better off with a newer drug, mefloquine (Lariam). The dosage is a single 250 mg tablet, one week before leaving, then weekly while in the malaria area and for 4 weeks after your return. Pregnant women should avoid Lariam because of possible adverse effects in the fetus, as should persons with epilepsy, seizures, or other neurological problems because this drug may have adverse effects on the nervous system. Alternatives to Lariam are chloroquine, the old standby (if you're *certain* that the area to which you're going has no chloroquine-resistant organisms), or 100 mg of doxycycline every day while you're in the malaria zone, continued for 6 weeks after getting home.

In addition to prophylactic antimalaria drugs, when in a malaria zone you should use insect repellent on your clothes and body (see page 158), and set up a mosquito-proof netting around your bed.

To determine what strain of malaria is present in the particular area you're going to, and to what drugs it is resistant, call the Centers for Disease Control Malaria Hotline at (404) 332–4555; it's open 24 hours a day.

Here's an example of how you can become sick with malaria when you least expect it. One of my colleagues, a professor of medicine at Cornell who heads a New York State Task Force on AIDS, recently visited Haiti to study the epidemiology of that disease there. On the plane coming home he developed shaking chills and a high fever, which he assumed were caused by either a urinary or respiratory tract infection. Malaria was the furthest thing from his mind. After a few hours, the fever and shakes suddenly disappeared—without any treatment! He was puzzled by this strange attack, and dismissed it as probably due to some unusual virus. A couple of days later, while lecturing at a meeting in New York, he suddenly developed another paroxysm of shaking chills and fever. It was so severe, he was unable to continue his presentation. Being a Cornell man, he finally made the right diagnosis of malaria, which was confirmed by a blood test. He was successfully treated and has had no recurrence since. This story emphasizes the point that there are areas in *this* hemisphere where malaria is endemic, a fact not always fully appreciated by most Americans or all professors of medicine!

There have actually been more than 1,000 cases of malaria diagnosed in the mainland United States in the last couple of years, most of them in California. The incidence is rising because of the large number of persons who harbor the parasite coming here from parts of the world where malaria is prevalent.

In addition to mosquito bites, you can contract malaria from a blood transfusion if the donor was infected and unaware of it. That's entirely possible because it may take as long as a year after being bitten by the mosquito before symptoms appear. Routine testing for malaria may not detect it unless the subject is actually having symptoms while the blood is being drawn.

Remember, persons with malaria feel perfectly well between paroxysms of fever. Sir Walter Raleigh, after he was sentenced to be beheaded, was very much concerned lest a malaria attack come on while he mounted the executioner's block. He did not want his enemies to think he was trembling with fear! If you suddenly develop shaking chills and fever for no apparent reason, see your doctor. Although a urinary tract infection is a much more likely cause, malaria is also a possibility, especially if you've been traveling somewhere exotic.

Once the diagnosis of malaria is made, the best treatment is the same drug used for its prevention, 1,250 mg of mefloquine (Lariam) in a single oral dose taken with 8 ounces or more of water. Such therapy, however, should be prescribed and administered by an expert in tropical disease medicine.

MALIGNANT MELANOMA: Fatal—But Curable!

Skin cancers, the most common malignancies of man, are not usually a threat to life. Left alone, most of them simply get bigger and bigger, and deeper and deeper. In the end, they may disfigure and become difficult to eradicate, but they rarely kill. An important exception to this rather benign scenario is the *malignant melanoma,* which, if untreated, is fatal.

As with so many other malignancies, the key to surviving a malignant melanoma is early detection, followed by a biopsy to determine to what extent it has penetrated the skin, and *immediate* removal. Caught in time, *this cancer is curable*. So, it behooves you to have a dermatologist take a close look at any doubtful skin "blemish," new or old. This is especially important if any of your blood relatives have had a melanoma, or if you are redheaded or fair-skinned. If you're lucky, your doctor may detect and remove a growth that is on its way to becoming a melanoma, but has not yet gotten there—a condition called *dysplastic nevus*. Here, too, excision means cure.

If the dermatologist thinks he has found a melanoma on your body but isn't sure, he will perform not only a simple biopsy, but a biopsy *removal*. In other words, he or she won't just take a small piece of the suspicious area to look at under the microscope and then go after the rest if it turns out to be a melanoma. Cutting into such a cancer can spread it; the whole thing should be removed in one sitting. If it is indeed a malignant melanoma, then if necessary the surgical margins can later be widened and the cut deepened in order to make sure that all of it was removed.

Is the "complete" (as far as one can tell) removal of a melanoma enough? Can you be certain that some microscopic cells haven't been left behind, especially since melanomas have the sneaky habit of reappearing 5 or 10 years after they were presumably totally excised? Should anything further be done just in case? There is no agreement on this vital question at the present time, but newer forms of treatment are currently being evaluated. Alfa interferon is the most promising to date, but other new approaches may also prove to be effective. For example, a vaccine made from your own melanoma is reinjected in order to make your body produce antibodies that will kill any cells that may have been left behind. And most exciting of all, researchers have already scored some success with gene therapy on this cancer.

For a day-to-day update of the whole melanoma story (and of other malignancies as well), call the Cancer Information Center at 1-800-4-CANCER. This information clearinghouse of the National Cancer Institute provides the latest melanoma news 5 days a week. They will, I am sure, also tell you that the key to melanoma control is prevention. That means *avoiding excessive exposure to sun*. As far as treatment is concerned, good and thorough surgery alone—the deepest and widest removal of the area involved—yields the best results, although it does leave you with an almost 10 percent risk of recurrence. If the cancer is located where a scar might be disfiguring, the operation should be done by a cosmetic surgeon experienced not only in beautification techniques, but in cancer removal too.

MEASLES: You Should Have Been Vaccinated!

Unlike its Teutonic namesake, German measles, the real measles (rubeola) makes you sick and miserable. Happily, the live vaccine developed in 1963 has been 99 percent effective in preventing this disease. So there is no excuse for anyone born after 1963 to contract measles.

Measles, like chicken pox, is very "catching" and easy to recognize. The first symptoms are fever, cough, and photophobia (when bright lights hurt the eyes), all followed by the characteristic Koplik spots on the skin—a red ring surrounding a tiny white center. The rash usually starts in the mouth and spreads from the head and neck to the rest of the body and the limbs, a sequence unique to measles. By the time the rash has reached the trunk, the spots on the face have begun to fade.

The treatment for measles doesn't cure the disease, but it makes the patient feel better. Pneumonia is one of the major risks of this infection, both in adults and in children. Unlike the pneumonia of chicken pox, which requires the antiviral agent acyclovir (Zovirax), this one responds to antibiotics, the most effective being amoxicillin (Augmentin) and cefaclor (Ceclor). However, these agents only *treat* the pneumonia that sometimes complicates measles; they do not prevent it. So someone with uncomplicated measles should not be given these antibiotics in anticipation, since they can actually increase vulnerability to *other* infections. Severe earache and encephalitis are additional complications of measles, and like the pneumonia, also respond well to Augmentin and Ceclor. Take Tylenol for the fever. (Avoid aspirin because of the risk of Reye's Syndrome, a severe neurological disorder.)

Here's an important new observation regarding measles therapy. Supplementing the diet of infants below the age of twelve months with 100,000 international units of vitamin A daily and that of older children with 200,000 units a day decreases the risk of death and disability from this infection, especially if there was a deficiency of that vitamin to begin with.

At age fifteen months, every child should be given the measles vaccine at the same time as he or she is vaccinated against the mumps, diphtheria, and German measles. Since this is a live vaccine, you should not receive it if your immune system is compromised, as, for example, by receiving immunosuppressive drugs for any reason. Persons taking cortisone should not be vaccinated either. Finally, if you were given a blood transfusion or a gamma globulin shot in the preceding three months, they may inactivate the vaccine —so hold off. If you're susceptible or were given the *killed* viral preparation used before 1963, you should be revaccinated with the more recent live preparation. A simple blood test will reveal whether or not you are immune.

We used to think that one shot of the measles vaccine at 15 months was enough. It appears, however, that a second shot should be administered upon school entry, and yet a third at approximately age 11 or 12. The reasons for the additional two shots is the apparent decrease in immunity with time even after this live vaccine.

MENOPAUSE: The Pause That Needn't Depress

Humans are the only mammals with a menopause. In every other species the female continues to menstruate throughout her entire lifespan. But, by the time a woman reaches fifty (earlier in some cases, later in others), her menstrual periods begin to taper off until they finally stop. Since the life expectancy of American women is now pushing eighty years, one third of our female population is postmenopausal.

Menopause occurs when the ovaries sharply reduce their production of female hormone (estrogen). That can happen abruptly when the ovaries have been removed surgically, or gradually, as part of the aging process. In some women, the cutback in hormone production has no major apparent impact aside from the cessation of the periods. But for many others, life changes dramatically. They become depressed, they suffer from hot flushes and flashes, their skin loses its moisture, the vaginal walls dry out, the bones become brittle and break as a result of calcium loss, and they are now as vulnerable as men to heart disease.

Should anything be done to correct this natural decrease in the estrogen level? Is there any risk in replacing the missing estrogen? Despite a great deal of research, the answers to these questions are still controversial. Given the same facts and statistics, the "naturalists" insist that menopause is nature's "will" and warn against meddling with it. Others, like myself, cannot fathom any conceivable benefit from estrogen depletion, and believe that replacing it not only enhances a woman's well-being, but also protects her against heart disease and osteoporosis.

If you are currently trying to decide whether or not to accept *estrogen replacement therapy (ERT)*, you may be interested in my recommendations:

I prescribe estrogens to my own menopausal patients (unless there is some overriding reason for them not to take them), especially those women

- who are white,
- who are thin,
- who drink,
- who earlier in life exercised so vigorously that they often missed their periods because they were already low in estrogen at the time,
- who took too little calcium in their diet over the years while trying to lose weight or lower their cholesterol levels by avoiding dairy products,
- who smoke,
- who never exercised enough.

When should estrogen replacement be started? How long should it be continued? What's the best way to take it—orally, by injection, or via skin patch? Who should *not* be using it? Are there any risks to this therapy, and if so, how can they be reduced or eliminated?

The two most feared consequences of ERT are cancer of the uterus and cancer of the breast. But the truth is, the likelihood of developing cancer of the lining of the uterus (endometrial cancer) is *not* increased if another hormone, progestin, is given along with the estrogen. In fact, most gynecologists now no longer think it is really necessary for women on estrogen to have annual endometrial biopsies if they're also on progestin and have not had any "breakthrough bleeding." Furthermore, endometrial cancer is not the kind that spreads wildly. If you see your gynecologist regularly once or twice a year, he or she can detect such a cancer early enough to cure it surgically.

With respect to estrogen and breast cancer, the jury is still out. My own interpretation of the available data is that ERT *probably does not* increase that risk, but I'm not absolutely sure. It may conceivably accelerate the growth of a malignancy *already* present, one so small that it has escaped detection on clinical examination. For that reason I suggest that every woman starting ERT have a screening mammogram to be certain that such a pre-existing cancer has not been overlooked. Even if the mammogram is normal, continue to examine your breasts regularly and repeat the mammogram every year (good advice even if you're not taking estrogens).

What's the best way to provide supplemental estrogen? If you have any history of blood clots or liver disease, use the skin patch. The estrogen it contains bypasses the liver and so does not influence blood clotting or any of its other functions. I also recommend the patch for women with hypertension since oral estrogens occasionally raise blood pressure.

The patch comes in 0.025, 0.05, and 0.1 mg strengths. Start with the 0.05 dose (equivalent to 0.625 mg of the oral form) twice a week for the first 25 days of the month, then stop for 5 or 6 days. If you develop a headache, or your breasts become tender, drop down to the 0.025 strength. If you experience "breakthrough bleeding," use the patch continuously without the 5 day abstention. If your uterus is intact, add oral progestin in the form of Provera, 10 mg a day for the last 10 to 14 days of the month. If you've had a hysterectomy, you don't need the progestin.

Estrogen supplements do slightly increase the chance of your developing gallstones, especially if there's a family history of this disorder, if you've had rheumatic fever, or if your cholesterol is high. They also occasionally cause migraine headaches, and make the breasts feel heavy and tender. However, the most annoying side effect is the return of the "period" with which you thought you were finally finished. All these complications are relatively uncommon, and are a small price to pay for the many benefits of ERT—the sense of well-being, the more youthful appearance, less thinning of the skin, prevention of vaginal dryness, retention of sexual desire, reduction in the incidence and severity of osteoporosis, and a 50 percent drop in the risk of a heart attack.

Estrogen replacement therapy should begin with the menopause and be continued for about ten years. I don't think it's necessary or beneficial beyond the seventies. But a word of caution about the accompanying progestin therapy. If you have coronary artery disease, or if your cholesterol is high with too much of the "bad" fraction (LDL) and not enough of the "good" kind (HDL), have your blood fats monitored at least twice a year, since progestin can further aggravate an abnormal blood picture.

In summary, here is the bottom line. I believe you should have estrogen replacement therapy beginning at menopause unless there is some special reason for you not to do so. If you still have a uterus, take progestin for 10 days a month in order to prevent uterine cancer. Every woman on ERT should have frequent routine breast examinations and annual mammograms because of a possible though unproven link between estrogens and breast cancer.

MENSTRUAL SYMPTOMS: They're Not All in the Head!

Some women sail through the entire menstrual cycle without any trouble, while others really feel lousy during some particular segment of their "monthlies." One difficult time for many is the actual period itself. We used to think that the pain, cramping, nausea, headache, diarrhea, and "blahs" associated with menstruation were all psychological. We now know better. They are, in fact, due not to the female personality but to a substance called *prostaglandin,* which makes the muscles of the uterus contract more vigorously. In fact, in order to induce labor, obstetricians administer a prostaglandin drug to get the uterus to squeeze hard enough to expel its contents.

The best treatment for menstrual pain is something that will block the production and/or action of prostaglandins. There are several such preparations that do just that, and they're all effective. The one that I recommend is ibuprofen, an NSAID (a nonsteroidal anti-inflammatory drug) which is marketed under a variety of trade names—Motrin, Advil, Nuprin, Medeprin, and several others. Most women require 400 to 600 mg every 6 hours for relief. You can buy the 200 mg strength without a prescription. If ibuprofen isn't effective, try a different NSAID. They vary enough so that naproxen (Naprosyn) or diclofenac (Voltaren)—these need a prescription—may be effective when ibuprofen is not (and vice versa). The dosage of Naprosyn is 375 mg, 3 times a day; for Voltaren it is 50 mg, 3 times a day after meals. Although the NSAIDs do control pain, they may also increase menstrual bleeding, so don't be alarmed if your flow pattern is somewhat altered when you're on one of these agents. If the NSAIDs don't work, mefenamic acid (Ponstell), 250 mg every 6 hours, may. It reduces the amount of prostaglandin produced, and *interferes with its action* on the uterus.

A word of caution. Don't assume that all the pelvic pain during your period is due to prostaglandins. In some women it results from endometriosis (see page 94), pelvic inflammatory disease, or an underlying structural problem in the pelvis. For example, the opening of the cervix may be so small that the blood shed during the period is not completely expelled. The uterus then contracts more and more vigorously in order to rid itself of the menstrual "waste." Or, the uterus may contract painfully because it contains fibroids. Finally, intrauterine devices (IUDs) occasionally worsen menstrual pain, too. If you are wearing one and your periods hurt, try a different method of contraception.

MIGRAINE HEADACHE: A New Treatment Allows You to Have Sex *and* Wine!

Everyone gets headaches now and then for a variety of reasons. The most common types are those due to stress, medication, infected sinuses, high blood pressure, a hangover, eye strain, a brain tumor, a stroke, a head injury, drinking red wine—or *migraine.*

Migraine headache is second only to tension headache in frequency. It affects more than eight million Americans, and is probably due to abnormal spasm and relaxation of the blood vessels in the head. We think something hormonal makes the vessels dilate and constrict because women are much more commonly affected than men; younger females taking oral contraceptives occasionally develop migraine, which may persist even after the Pill is stopped; these headaches are frequently worse during menstrual periods; they often ease up with pregnancy, and sometimes end with the menopause. But migraine affects both sexes, and may also clearly result from food and drink as well as from emotional stress or hormonal fluctuations.

The management of migraine headaches involves both prevention and treatment. For example, if you know from experience that you're surely going to develop one after eating chocolate, aged cheese, hot dogs, Chinese food containing MSG, shellfish, too much coffee, or certain wines, then the best treatment is to avoid that specific dish or drink. However, in most cases, migraine comes on out of the blue—without any apparent reason.

In classic migraine, the headache is one-sided, and is accompanied by nausea and vomiting; it may be preceded by an hour or more of other symptoms, like seeing flashing lights, spots, or zig-zag lines, or by strange odors. If you are "lucky" enough to have such advance warning, you may be able to prevent the full-blown attack with 2 tablets of isometheptene (Midrin) at the very first intimation of trouble, and 2 more an hour later. This drug constricts the dilated cranial blood vessels responsible for the migraine, while the acetaminophen and sedative it contains help relax you. Ergotamine (Ergomar) will also abort migraine by narrowing the dilated vessels, but since it often causes nausea and gastric upset, I prefer Midrin. Avoid Ergomar if you have angina or are pregnant.

If neither Midrin nor Ergomar prevents the headaches, or if yours is the kind that comes on suddenly without any warning, then for pain relief I suggest you try 1 Fiorinal tablet (a combination of a barbiturate and aspirin) every 4 to 6 hours, or naproxen (Anaprox), a nonsteroidal anti-inflammatory drug that seems to be more effective against migraine than are other agents in this group. In some people the headache can be so severe as to require narcotics like codeine and occasionally even meperidine (Demerol). A drug that may represent a major advance in migraine pain control was developed in 1991. It is called sumatriptan; it constricts the arteries that are dilated in the acute attack. One injection into the arm results in quick and dramatic pain relief without significant side effects.

Here is some more good news about migraine. It seems that the more sexually active you are, the fewer headaches you're going to develop. So, the "No, dear, not tonight" excuse should be replaced by the "Please, dear, tonight" request! More important is the observation that 1 aspirin *every other day* reduces the incidence of recurrence by about 20 percent. For patients who suffer more than three migraines a month, and in whom aspirin is not effective, I prescribe the beta-blocker propranolol (Inderal) on an ongoing basis in a dosage of up to 240 mg a day. However, this drug should not be taken by insulin-dependent diabetics or those with heart failure, chronic lung disease, or allergic asthma. If Inderal does not reduce the frequency of attacks, I advise diltiazem (Cardizem), a calcium channel blocker, one 90 mg sustained-release tablet a day.

There is something in *red* wine that gives certain vulnerable individuals a headache. It's not the alcohol (because these same persons tolerate white wine without any problem), nor is it sulfites (also present in white wine). Whatever does it, a very unpleasant headache follows even small amounts of red wine. Do you miss being able to drink an exciting bordeaux, or a full-bodied burgundy? Here's some great news! If you take 2 aspirins just before imbibing, chances are you'll be headache-free—and don't ask me why!

In summary, some migraine headaches give advance warning that permits you to take an agent to abort the full-blown attack. The two most effective ones are Midrin and Ergomar. Once the headache starts, symptoms often require potent pain killers. The frequency of migraine recurrences can be reduced by a variety of agents, the simplest and most convenient of which is an aspirin every other day, or if that doesn't work, a beta-blocker or calcium channel blocker.

MITRAL VALVE PROLAPSE: All About Clicks, Murmurs, Dentists, and Antibiotics

For years doctors ignored the "innocent" murmurs and strange little clicks they sometimes heard when listening to a patient's heart. These "peculiar" sounds, for which there was no explanation, were usually audible in thin, nervous, but otherwise healthy females. The rest of their physical exam was apt to be normal as were the electrocardiogram and the chest X ray. Then in 1963, an astute South African physician, Dr. John Barlow, suggested that these cardiac sounds were due to a billowing or ballooning of portions of the mitral valve in the heart whenever it closed. He named this condition floppy mitral valve, but it is now more commonly referred to as mitral valve prolapse (MVP).

With the invention of the *echocardiogram,* it became possible for the first time to observe the movement of all the structures within the beating heart. In persons with a click, one could actually see the mitral valve billow when it closed, just as Barlow said it did. Sometimes, when the cusps of the valve did not shut tightly, this billowing was accompanied by a leak of blood across the valve. Those two phenomena—the *billowing* and the *leak*—are the hallmarks of mitral valve prolapse. What do these findings mean? As long as the valve is not leaking substantially, the billowing and the accompanying click are harmless (the click itself is simply the sound the valve makes after the billowing motion is completed). However, the presence of a murmur in addition to the click means that there is some leak of blood across the mitral valve, in which event the MVP assumes more importance. In any case, it rarely interferes with the quality or duration of life.

The echocardiogram has been a mixed blessing as far as MVP is concerned. On the one hand, it has explained certain puzzling findings and symptoms, but unfortunately, many healthy individuals have come to believe that they have heart disease simply because their mitral valve is "prolapsed"—a term that may not sound ominous when applied to the rectum or vagina, but is frightening in a cardiac context.

Even though mitral valve prolapse is, on the whole, a benign condition, certain precautions and/or treatment are sometimes necessary. The most important of these is the use of prophylactic antibiotics whenever a surgical or invasive diagnostic procedure is performed—such as a D&C, cystoscopy, colonoscopy—as well as at virtually *every* visit to the dentist no matter what he does to you. Unless you take those antibiotics, you run the risk of infecting the prolapsed mitral valve, a condition known as *subacute bacterial endocarditis (SBE)*. Before the availability of antibiotics, SBE was usually fatal. Although it can now be cured, it is much better avoided. New evidence suggests that if you have only a click *without a murmur,* these antibiotics may *not* be necessary. So ask your doctor whether your mitral valve prolapse is generating a murmur as well as a click.

The best antibiotic to protect the prolapsed mitral valve (unless you are allergic to penicillin) is amoxicillin, a distant penicillin relative. Take 6 tablets, 500 mg strength, about an hour before the dentist begins working on you, and 3 more 6 hours later. If you cannot tolerate amoxicillin, you may use erythromycin, 4 tablets of the 500 mg strength 2 hours before the dental (or other invasive) procedure, and 2 tablets (1,000 mg) 6 hours later. Remember that cleaning the teeth is as much a dental procedure as a filling or an extraction as far as your prolapsed mitral valve is concerned. When the gums bleed for whatever reason, the bacteria that normally inhabit everybody's mouth get into the bloodstream and can settle on and infect the prolapsed valve. In fact, 20 percent of all cases of SBE occur in persons with mitral valve prolapse who failed to take the appropriate antibiotics.

Another possible complication of mitral valve prolapse in which there is a leak across the valve is gradual cardiac enlargement and heart failure that may ultimately make repair or replacement of the valve itself necessary. You should suspect that this is happening if you become progressively more short of breath. However, that has happened in only a handful of the many hundreds of people in whom I have diagnosed MVP.

If you have a prolapsed mitral valve, you should know this:

- It affects as many as 6 percent of otherwise perfectly normal women, and about half that number of men.
- It rarely constitutes a threat to life—but it can.
- Individuals with MVP almost always lead completely normal lives—and that includes having as many children as they want, and exercising to their heart's content.

Symptoms of Mitral Valve Prolapse and Their Treatment

Although most persons with MVP have no symptoms (at least until they're told about their condition and begin to worry about it), some do have complaints that may require treatment. These include an irregular heartbeat and palpitations (a pounding or awareness of the heart's action), migraine headaches, chest pain, and "panic attacks." Here's the best treatment for each of these complaints:

• For an *irregular heartbeat,* even if it's annoying or frightening, reassurance is usually all that's necessary since most of these rhythm "abnormalities" are harmless. However, in a very few such individuals, the arrhythmia can result in cardiac arrest or a stroke and *must* be treated. Under these circumstances, or when a patient is too uncomfortable or worried, I prescribe a beta-blocker like nadolol (Corgard), 40 mg once a day.

• The best therapy for the immediate relief of *migraine headaches* associated with MVP is Fiorinal—a combination of the barbiturate butalbital and aspirin—1 tablet or capsule every 4 to 6 hours. Don't use this drug any more than is necessary because it fosters dependence. If the migraines recur frequently, an aspirin every other day, or a beta-blocker like Inderal, or a calcium channel block such as diltiazem (Cardizem) (see page 139) may help prevent them.

• For some unknown reason, many patients with mitral valve prolapse complain of *chest pain.* MVP-associated chest pain is *not* usually due to coronary artery disease, and *never* results in a heart attack. If aspirin, Tylenol, or an NSAID (one of the aspirin-like nonsteroidal anti-inflammatory drugs) is not effective, 10 mg of Inderal may be.

• *Panic attacks*—sudden extreme anxiety for no apparent reason—also commonly occur in persons with MVP. They are probably due to an overproduction or release of adrenaline in the body. The best treatment for them is imipramine (Tofranil), 25 mg, 3 times a day. Although alprazolam (Xanax), 0.5 to 1 mg, 3 or 4 times a day is also effective, I worry about the possibility of habituation with this agent. If after taking either of these medications regularly for 2 weeks or more you no longer need them, taper the dosage over a 10-to-14-day period; do not stop them abruptly.

In summary, the vast majority of cases of MVP are uncomplicated, and permit you to enjoy a completely normal life in every way. But some individuals do experience symptoms that may require treatment and, rarely, even replacement of the valve itself. Antibiotics are usually necessary to prevent infection of the valve after surgical procedures or dental work.

MUMPS: And Your Sex Life

A forty-five-year-old man called to tell me the following: Several weeks earlier his seven-year-old child had come down with mumps, from which she had recovered completely. A few days ago my patient noticed that his own salivary glands (situated in the cheek below each ear) were swollen. Aside from looking like a chipmunk, he had no special complaints. Now, two or three days later, he awakened to find his scrotum swollen, tender, and red. He felt sick to his stomach, vomited a couple of times, and had a temperature of 100 degrees. He phoned for an urgent appointment because a friend had panicked him by predicting that since the mumps had spread to his testes, he would surely be sterile and probably impotent forever! The threat of sterility didn't worry him; he already had three kids, and in fact, had been giving serious thought to a vasectomy to prevent a fourth. But impotence! At forty-five! That terrified him!

When I examined him later that day, I confirmed that he did have mumps and that the disease had in fact spread to his testes. This happens in male adults in one third of cases. But even when it does, it *never causes impotence,* and the incidence of sterility is very low—less than 2 percent! So I reassured him and prescribed the following treatment: Codeine for the pain, a good jock strap to support the testes, and ice to the tender area. He called 3 or 4 days later to tell me he was feeling well and that *everything* was "working" just fine. Had his testicular pain persisted, I'd have prescribed a steroid by mouth for about a week. There is a postscript to this particular story. Six months later I received yet another phone call from this gentleman, this time to inform me that his potency and fertility were both intact. His wife was expecting their fourth child!

Mumps is a viral disease of childhood for which there is an effective and safe vaccine. It's easy to recognize by the painful, swollen parotid (salivary) glands that leave one looking like a chipmunk. Since the virus is transmitted in the saliva, such patients should avoid kissing anyone who is not certain that he either had the mumps or was vaccinated.

Mumps lasts for a few days and is usually without complications. The *best treatment* is to apply moist packs, cold or warm, to the swollen glands, and to shun sour or spicy foods, which make the pain worse by stimulating the glands. Use acetaminophen (Tylenol) 4 or 5 times a day as necessary to lower an elevated temperature and to control pain. You can always move on to codeine if the Tylenol doesn't work. Remember, no aspirin for this and other viral illnesses in children because of the risk of a serious neurological disorder called Reye's Syndrome.

Although the great majority of mumps cases clear up without any problems, about 10 percent of patients do develop severe headache and other signs of meningitis (irritation of the brain covering) from which they recover in about a week. Other complications such as deafness, inflammation of the pancreas, and encephalitis (inflammation of the brain itself rather than its coverings) can also occur, but are extremely rare.

Although mumps vaccine is safe (almost 80 million doses have been administered in this country alone in the last 25 years), since it is a *live* vaccine, *it should not be given to pregnant women,* to persons receiving cancer chemotherapy whose immunity may be impaired, or to individuals allergic to chicken, chicken feathers, or eggs (the vaccine is prepared in egg cultures).

Doctors usually vaccinate children in the second year of life as part of the routine immunization schedule as well as persons born *after 1957* who have not had mumps. (If you're older than that, you're almost certainly immune.) If there's any question, get the vaccine.

NAIL DISORDERS: Infected, Distorted, Brittle, and Breakable

The other day while walking along a Florida beach, I shifted my gaze from more interesting parts of the sun worshippers to their feet. I was horrified to see how deformed many of them were—toes overlapping, squished together, and pointing in every which way. The nails, too, left much to be desired esthetically—misshapen, gnarled, and discolored. But the fingers in the very same people were, by contrast, usually "normal" except for the occasional case of arthritis.

Toes become distorted because of the tight, badly fitting shoes we wear in the name of fashion. Those high heels and the narrow, pointed shapes may look sexy, but they were not designed with nature in mind. But even when the footwear is sensible, moisture that accumulates under one's socks provides fertile feeding for fungi that turn healthy, pink toenails into yellow or dark brown "uglies."

• *Fungal nail infections* may be difficult to treat, especially when they involve the big toe. Once the fungus settles under the nail bed or comes in from the sides, there's nothing you can apply to the surface of the nail itself that will help. You now need an *oral* antifungal preparation. Griseofulvin (Fulvicin), 1 500 mg tablet taken twice a day with meals, is usually effective and safe, but if it doesn't clear the infection, ketoconazole (Nizoral), a newer and more powerful antifungal agent, will. The dosage of the latter is 1 or 2 of the 200 mg tablets daily, usually continued for months. Ketoconazole is a great drug for *serious* conditions, but it can hurt the liver. So make sure that your liver function is normal *before* you take a single tablet, and repeat the blood tests every 3 or 4 weeks while you are on this drug. Take ketoconazole only when necessary, and be sure to stop it as soon as it has done the job. Also, never use it for prevention, only for treatment.

Once healthy-looking areas begin to appear on the nail, you may then apply a *topical* antifungal preparation and continue to do so until the nail is fully restored. That may take as long as a year because nails only grow 0.1 mm a day on the fingers, and even more slowly on the toes.

• An *ingrown toenail,* unlike fungal infections, is not an eyesore, but hurts like the devil. You can usually lift it out yourself, but sometimes when the nail is deeply imbedded in the tissue and also infected, its removal may require the services of a doctor or podiatrist. To prevent recurrences, wear shoes that don't squeeze your toes together, and always cut your nails straight across along the top. Occasionally, a portion of the nail and root needs to be excised, in which event, it doesn't grow back. That happened to me when I was five years old, before the era of antibiotics. For some reason, my right great toenail kept growing sideways rather than straight ahead, and was constantly digging into the tissues and infecting them. In desperation my family doctor cut away half the nail down to its roots to cure the problem. That took care of it, but the nail never grew back. If you're ever on a beach and see a tall, thin, good-looking man walking around with half a toenail on his right foot, come over and say hello to me.

• Do you have *brittle fingernails* and do they break or split easily? While the problem may be due to infections, injury, psoriasis, or some "internal" disease, it is more likely the result of chronic exposure to moisture or to nail polish. The best way to deal with brittle nails when the cause is obviously cosmetic is to remove nail polish with oily solvents rather than those with an alcohol base. Keep your nails short. Clip them only after you've softened them in a bath or shower, not when they're dry and wear gloves when you're working with water or any chemicals.

My wife insists that I add the following piece of advice. In *her* experience, anyone whose nails keep splitting should try "Fabu-Nail" by Revlon. She uses it instead of nail polish. It's colorless and she swears that it prevents her nails from breaking. You might as well try it or I won't hear the end of it. But remember, this is her suggestion, not mine. Finally, I have never been able to substantiate the popular notion that gelatin capsules prevent nails from breaking.

Tell your family doctor about your nail problem to make sure it isn't the result of undetected diabetes, iron or vitamin deficiency, or an underfunctioning thyroid. Correcting the underlying abnormality will restore the nails to their original pristine beauty.

- *Biting, picking at, and tearing of the fingernails* is a nervous habit and not a disease. Even though it leaves the nails with a telltale look and less than attractive, it rarely results in a visit to the doctor's office. The only "treatment" is a tranquilizer, which I don't think is worth taking. In my opinion, you're better off biting your nails than taking a psychotropic drug.
- If you are a nail biter, chances are you also have *hangnails.* If you do, cut them away with a sharp pointed scissors before they become infected.

NAUSEA AND VOMITING: The Symptoms Are in Your Stomach, But the Decisions Are Made by the Brain!

The more I deal with the body in health and disease, the greater my awe at how beautifully all its organs and systems interact. Take nausea and vomiting, for example. You experience the cramps and most of the other "ugh" feelings in the stomach, but the *decision* to vomit is made far away by the vomiting center in the brain. There, tiny cells receive and interpret a slew of signals on the basis of which they decide whether or not you should vomit. Those messages may come from the stomach (when you've been drinking too much, eaten tainted food, or been poisoned); from the gallbladder (indicating that it is inflamed, infected, or full of stones); from the liver (signaling the presence of hepatitis); from the kidney (warning that it's not working properly and is retaining toxic substances); from the labyrinths in the inner ear (whose malfunction makes you seasick); or from other parts of the brain itself (when you see something abhorrent, repulsive, or smell a foul odor).

So to control nausea and vomiting, you must eliminate the stimuli that are activating the vomit center—cut down or stop drinking, be more careful about what you eat, have the diseased gallbladder removed (but make absolutely sure that that's what's giving you the symptoms), treat the liver disease, correct your kidney function, or reduce the dose of whatever drug is making you sick (digitalis is the prime example of such an agent). When these options are not possible (your kidney disease may be irreversible, or there is no effective treatment for your hepatitis), then you can and should override or "numb" the functions of the vomiting center. Nausea and vomiting are beneficial when you've swallowed a poison, but once you've emptied your stomach, there's no point in continuing to retch. Unfortunately, the vomiting center doesn't always exercise the best "judgment." For example, it may force you to throw up a medication that you actually need, such as an effective anticancer drug or a necessary antibiotic. Were it not for our ability to "inactivate" the vomit center, we would be forced to discontinue several life-prolonging and life-saving agents.

The first step in treating nausea and vomiting is to *replace the fluids lost,* just as you would when you have diarrhea. This is especially important in the chronically sick, the elderly, or children. When adequate fluid replacement is not possible by mouth because even sipping water makes you retch, you may require intravenous fluids.

Some of the most common causes of nausea and vomiting are motion sickness, pregnancy ("morning sickness"), gastroenteritis, anticancer drugs, pain killers (especially narcotics), psychogenic vomiting, and ear problems.

Motion Sickness

Motion sickness can affect anyone. *Antihistamines* have traditionally been the most widely used agents for its prevention and treatment, and I still prefer them for short plane or car trips or ferry rides across a stormy bay. The product I like best is cyclizine (Marezine), which comes in a 50 mg strength and is available over the counter. Take 1 about 30 minutes to an hour before setting out, and every 4 hours thereafter as necessary (except if you're driving the car yourself!). I'm not aware of antihistamines ever having harmed a human fetus, but since they can result in congenital abnormalities in rats, don't take them if you're pregnant. An anti-motion sickness drug, *cinnarizine,* currently available only in Europe (and without a prescription at that!), is in my opinion more effective than anything we now have in the United States. Perhaps by the time this book is in the stores, the FDA will have approved it. When it does, that's the one to use.

I no longer recommend antihistamines for long cruises or ocean voyages because they so often leave you sleepy and drowsy. What's the point of spending thousands of dollars on a "fun" trip only to be confined to your room? Here, I advise a skin patch impregnated with scopolamine. This preparation, marketed as Transderm Scōp, has been around for several years. Put it on from 4 to 12 hours before you set sail, but first wash and dry your hands thoroughly. Also, clean the area where you're going to place it to prevent the absorption of dirt or other impurities along with the medication itself. Although the manufacturer recommends affixing it behind the ear, I prefer the upper arm, knee, or elbow because the medication is less rapidly absorbed there and so results in fewer side effects. Scopolamine will protect you for about 72 hours. Again, although the package insert advises that you change the patch after 3 days, I suggest you remove it after only 2. Then wait 24 hours before putting on a fresh one. There is usually enough medication left in your system to protect you during that drug-free day, and I have found that the smaller dose reduces the severity of side effects due to "withdrawal." These consist of all the symptoms you tried to avoid on board ship, and they suddenly hit you later, on dry land ("seasickness ashore").

Although scopolamine is extremely effective, it may produce blurred vision, drowsiness, temporary confusion, a sense of giddiness, restlessness, and in men with enlarged prostates, trouble urinating. But in my experience, these adverse symptoms are relatively infrequent and usually more tolerable than the relentless nausea and retching of motion sickness, and should not deter you from using the patch. But scopolamine is for adults only. *Do not give it to children,* and avoid it if you have glaucoma, any urinary problems, or pyloric obstruction (an uncommon condition in which there is a narrowing between the stomach and the small intestine that causes paroxysms of vomiting). Also, even though there is no proof that scopolamine has any adverse effects on the fetus, I don't recommend it if you're pregnant or are nursing. In my view, no pregnant woman should ever take any optional drug.

Here's some practical advice about seasickness that won't necessarily replace the medications I've discussed, but will help. As soon as you become nauseated and dizzy, lie down. Don't look out the port hole. No matter how calm the sea, the horizon will always be moving up and down and that will worsen your symptoms. If you're determined to keep your eyes open, then just look at the solid wall straight ahead.

"Morning Sickness"

Some doctors think it's all in your head—the "morning sickness" that troubles as many as 80 percent of women during the first three or four months of pregnancy, but I don't believe it. I simply cannot accept the fact that these females *all* have virtually identical psychological problems at the same time of their lives—*all* of which disappear after the fourth month. Some hormonal fluctuation *must* be doing it. Although the nausea occurs most often in the morning, some women feel queasy all day. Frequent small meals rather than larger ones may help you feel better, but if you *must* have something more for relief, try 1 to 2 teaspoons of Emetrol every 3 to 4 hours. This is a phosphorated carbohydrate (medicalese for sugar) with no significant potential toxicity. Despite my favorable experience with this remedy, *you should know that the FDA has not approved it or any other* medication for the treatment of morning sickness.

Gastroenteritis

Gastroenteritis is a fancy word for an upset stomach. You're nauseated, you vomit, your belly hurts, and if you're not throwing up, you're burping. The cause of all this misery may be a virus, or something you ate or drank. It may even be that new medication your doctor prescribed.

If you have acute gastroenteritis, food is the furthest thing from your mind. But if you've been vomiting a great deal, dehydration can present a problem, and fluid replacement is important, especially in children, the elderly, and the chronically ill. So try to take as much liquid (ginger ale or consommé) as you can comfortably keep down. Once you feel like eating again, even just a little, begin with small portions of something bland like boiled chicken, rice, and toast.

Acute gastroenteritis usually clears on its own within 24 hours. But if the nausea persists and prevents you from retaining any food and liquids, try a

250 mg trimethobenzamide (Tigan) suppository first. Emetrol, the phosphorated carbohydrate mentioned above, is equally good and not apt to hurt you. If neither Tigan nor Emetrol does the trick and your nausea continues, a prochlorperazine (Compazine) suppository is the next step. It comes in a 25 mg strength, and you can take up to 3 a day (there's no use trying the oral Compazine because chances are you're going to vomit that, too).

Anticancer Drugs

There are many thousands of people in this country who require chemotherapy to control, slow down, or cure a cancer. Most of these agents cause nausea and vomiting, and before we were able to control these symptoms, patients would tell me they preferred to die rather than endure the toxicity of the "cure." But things have changed. That is not to say that cancer chemotherapy is fun or easy, but the resulting nausea and vomiting can now largely be managed.

There is a phenomenon called *anticipatory emesis*—the anxiety based on the memory of having taken a particular medication, which in itself can produce nausea and vomiting, independent of the action of the drug. Combining a tranquilizer like lorazepam (Ativan), 1 mg, 3 or 4 times a day, and prochlorperazine (Compazine), 10 mg doses every 6 hours, will usually prevent this nausea and vomiting as well as that due to *mildly* emetogenic (vomit-inducing) anticancer agents such as methotrexate, 5-FU, cytarabine (Cytosar-U), and tamoxifen (Nolvadex). But if you're taking "heavy hitters" like nitrogen mustard, cisplatin, adriamycin, cyclophosphamide, CCNU, BCNU, and the like, you'll require something stronger than Compazine. I recommend metoclopramide (Reglan) *intravenously.* Your doctor, or a specially trained chemotherapy nurse, will administer it to you about a half hour before the actual anticancer drug is begun, and every 2 hours thereafter for about five or six doses. Reglan, which, incidentally, in smaller amounts orally is used in the treatment of reflux of food or acid from the stomach up into the esophagus, may result in symptoms of Parkinson's disease (tremor, rigidity of muscles). You can decrease the likelihood of that complication by taking 50 mg of diphenhydramine (Benadryl), an antihistamine, along with it. More recently, the FDA approved ondansetron (currently given intravenously in this country but orally in Canada) for the treatment of nausea caused by anticancer drugs. Initial reports say it is more effective for this purpose than Reglan. If you are receiving drugs that induce severe nausea and Reglan isn't working, ask about ondansetron.

Until quite recently, you would have had to go to the hospital to receive intravenous chemotherapy. But many doctors now administer it either in their offices or in out-patient clinics. You come in the morning, spend the day there, and go home in the evening.

In addition to the medications listed above, I also advise high doses of steroids like dexamethasone (Decadron) or methylprednisolone (Solu-Medrol) intravenously. This combination almost always reduces the nausea and vomiting to tolerable levels. When it doesn't, I recommend marijuana. Yes, *pot!* Don't be shocked. It's perfectly legal in these circumstances everywhere except Massachusetts. This otherwise tightly controlled drug has been approved by the FDA for the control of nausea in patients in whom conventional agents have not been effective. (Its brand name is Marinol.)

Pain Killers

Most narcotics taken for pain control can leave you nauseated—some more than others. The worst offender, and yet the most effective against pain, is morphine. Codeine is better tolerated, but is a weaker pain killer. If the analgesic you're taking causes nausea, prochlorperazine (Compazine), 10 mg every 6 hours, or promethazine (Phenergan), 25 mg, 3 or 4 times a day, will make you feel better as will the antihistamine meclizine (Antivert), 12½ mg, 3 or 4 times a day.

Psychogenic Vomiting

Psychogenic vomiting is an emotional disorder that affects mostly young women. It's probably a form of anorexia or bulimia. Such persons often throw up after eating even though they are not nauseated. They usually do so secretly, and never complain about it either, almost as if they were deriving subconscious gratification from being thin. Like anorexia, psychogenic vomiting is very difficult to treat. Antinauseants are no help. Until such time as we come to understand its cause, this problem (which, incidentally, sometimes runs in families) is best managed by a psychiatrist.

Ear Disorders

If you awaken one morning feeling perfectly well, but the room spins and you're overwhelmed by a feeling of nausea when you try to get out of bed, the diagnosis is almost certainly *viral labyrinthitis*. However, the sudden onset of dizziness and vertigo in older individuals, especially those with vascular disease, does raise the question of a stroke. If you fall into that vulnerable category, see your doctor as soon as possible. The best treatment for viral labrynthitis is to remain flat on your back and take an antihistamine like meclizine (Antivert), 12½ or 25 mg, 3 or 4 times a day. Viral labyrinthitis always clears up in a few days. If it doesn't, you should be evaluated by your doctor.

Other Diseases

Nausea and vomiting accompany a host of different diseases besides viral labyrinthitis and stroke, including migraine, peptic ulcers, and gallbladder trouble. Each requires specific treatment to resolve the underlying problem. But at the same time, you can use the various medications described above to minimize the common symptoms—nausea and vomiting.

NOISES IN YOUR HEAD (TINNITUS):

That No One Else Hears!

Millions of older people experience buzzing, ringing, hissing, or clicking noises—intermittent or permanent—in one or both ears. Doctors call this condition tinnitus.

By far, the most common cause of tinnitus is wax in the ear canal, in which case the symptom is quickly "cured" when the wax is removed. The noise will also disappear if it is the result of your having taken too much aspirin, certain antibiotics, or quinidine, a widely used cardiac drug. But in most other instances, the tinnitus is usually forever. It may be better some days and worse others, but it rarely disappears. Arteriosclerosis of the arteries in the ear and brain is the most frequent cause of permanent or nonreversible tinnitus. But long-standing exposure to loud noises at work (jack hammers) or at play (discos, blaring personal radios and cassette players) or a tumor involving a nerve within the ear (acoustic neuroma) may also be responsible. This latter diagnosis is an important one to make, because surgery not only cures the tinnitus, it is life-saving as well. If the reason for the sudden noises in your head cannot be determined in any other way, you should have either an MRI or a CAT scan.

Except for removing wax from the ear canal, eliminating medications that can cause it, or surgically removing an acoustic neuroma, there is no cure for tinnitus. Some doctors prescribe nicotinic acid and high-potency vitamins, which I have never found to help one bit. Your best bet is to buy a *tinnitus masker* at a surgical supply house or drug store. This is a little gadget that fits into the ear and creates a more pleasant noise of its own that masks the one that offends. You'll find it most effective in the silence of the night, when your native tinnitus is at its loudest.

OSTEOARTHRITIS: The Joints Ache But the Rest of You Feels Fine

Osteoarthritis, which most of us develop to some degree as we get older, is a disorder affecting weight-bearing joints like the feet, knees, hips, and back. Whereas rheumatoid arthritis has a predilection for women, osteoarthritis occurs with the same frequency in both sexes. Even though its pain can be considerable, unlike rheumatoid arthritis, there are few if any accompanying "systemic" symptoms such as fever, fatigue, and generalized malaise. Persons with osteoarthritis typically awaken in the morning with a stiff back, hip, or knee that improves as they limber up during the day. (By contrast, the stiffness in rheumatoid arthritis lasts for hours and, depending on the severity of the disease, can be continuous.)

The best treatment for osteoarthritis basically involves proper exercise and effective pain control. A good physiatrist or physiotherapist can prescribe and monitor an exercise program that will improve joint function and reduce pain. Moist heat 2 or 3 times a day, traction, hydrocolator packs, and similar devices can also help. Try as many as necessary to find which combination is best for you.

As far as pain management is concerned, I usually go directly to one of the NSAIDs (aspirin-like nonsteroidal anti-inflammatory drugs) (see page 311), which I prefer to aspirin because they are more effective and less irritating to the stomach. (But use Tylenol or Trilisate after meals if you have a peptic ulcer or any bleeding disorder.) Older individuals on NSAIDs should look out for kidney complications, especially if they also have high blood pressure or heart failure. My favorite NSAID for osteoarthritis is one of the over-the-counter preparations of ibuprofen (Medeprin, Advil, Nuprin, Motrin IB). If this 200 mg nonprescription strength is not enough, ask your doctor for a stronger one, my preference being diclofenac (Voltaren), 50 mg, or naproxen (Naprosyn), 375 mg, 3 times a day after meals.

An imaginative doctor who gives careful thought to your symptoms can come up with an innovative approach that may alleviate them. For example, I have one patient with very big breasts that were pulling her arthritic spine forward, leaving her very uncomfortable. Breast reduction surgery solved that problem. You may benefit from wearing sneakers or specially made shoes if there are osteoarthritic changes in your feet. Instruction regarding posture is important if your spine is arthritic, and you may have to change your work environment if your job requires standing for long periods of time.

If osteoarthritis involves your knees or hips or shoulders, and is getting worse so that you have lots of pain and can't move about comfortably, you now have a wonderful option—*joint replacement*. Don't be afraid of it. When done by a skilled surgeon, it will vastly improve the quality of your life.

OSTEOPOROSIS: Brittle Bones That Break

Osteoporosis is a "thinning" of the bones that occurs when the calcium that keeps them strong has seeped out. This disorder is responsible for the 1.3 million fractures suffered annually by older females. Occasionally just a sudden vigorous movement or forceful cough may cause one or more ribs to break. When the spinal column cracks (usually spontaneously), the involved vertebrae collapse and the woman loses some of her height and develops the "dowager's hump." Each year 300,000 elderly ladies fall, slip, or are mugged in this country and fracture a hip—an event that usually means hospitalization and surgery. One in four is never well enough to return home—most end up in a nursing facility, some remain crippled, and many die.

Aside from the pain and suffering, the complications of osteoporosis cost society an estimated $10 billion in medical care every year.

Although "brittle bones" affect both sexes, it's a much more common and serious problem in women after menopause because the process of depositing calcium in bone and keeping it there is very much dependent on the action and availability of female hormones. But nonhormonal factors play a role, too—*race* (black women are less vulnerable than are Orientals and Caucasians); inherited *body build* (if you are thin and petite, you're more likely to become osteoporotic than your big-boned, heavier friends); *too much alcohol* with its associated poor nutrition; heavy *cigarette smoking;* inadequate intake of *calcium* before menopause, when it really counts; lack of *exercise;* chronic use of *steroid hormones;* an *overactive thyroid* gland; tumors of the *parathyroid* glands, which secrete abnormal amounts of a hormone that sucks calcium out of the bone. All of these leave you vulnerable to osteoporosis, but the major cause is a diminished estrogen level after menopause.

Bearing all these risk factors in mind, there are several steps you can take to *prevent* and then, if necessary, to *treat* osteoporosis. *The key to prevention is depositing enough calcium in your bones before menopause* by eating a proper diet and engaging in regular exercise of the weight-bearing kind, like walking or jogging, at least three or four times a week. That renders the bony skeleton thick and strong enough to withstand the hormonal impact of menopause when the calcium deposits begin to seep out. It's very much like a bank account. You build it up while you can, so that you'll have enough to call upon in the lean years. Unfortunately, most women do not think about osteoporosis when they're young and vigorous. It's only as their periods taper and finally stop that they become calcium-conscious—and by that time, it may be too late.

In order to ensure a solid bony skeleton, every female should consume at least 1,000 mg of calcium a day beginning in her teens and increase it to 1,500 mg as she approaches menopause.

The richest *dietary* sources of calcium are dairy products (one ounce of cheese contains 200 mg) and meat—precisely what you'd be avoiding if you were following a low-cholesterol diet. In that case, focus on tofu, sardines, and leafy green vegetables. Most women on such a diet, however, should also be taking calcium supplements. Remember, however, that regardless of the calcium content listed on the bottle, some of these commercial preparations are better absorbed than others. To evaluate how much calcium you're actually getting from your present supplement, drop the tablet into 6 ounces of white vinegar, leave it there for half an hour, then stir gently. If it hasn't dissolved in the glass, it's not likely to do so in your stomach either.

In my book *Modern Prevention,* I advised taking 400 units of vitamin D along with the calcium supplement in order to increase the latter's absorption from the gut. That recommendation is no longer considered valid. (How quickly "facts" change in medicine!) Unless you're clearly deficient in vitamin D, do *not* take additional amounts with the calcium. You may end up getting more vitamin D than is good for you.

Although it's important to maintain adequate calcium intake *after* menopause, there are more effective additional measures to be taken. If you've done all the right things and reached menopause with your bones intact, you can reduce the risk of osteoporosis later on by taking *estrogen replacement therapy (ERT)* for about ten years after your periods have stopped (see page 184). And that's not the only health dividend to be derived from such replacement. You'll also experience a sense of well-being, and enjoy a 50 percent reduction in the risk of heart attack. However, do not take estrogens if you have undiagnosed vaginal bleeding, a blood-clotting problem, liver disease, or previous cancer, especially of the breast, ovaries, or uterus or if there is a strong family history of these malignancies.

If you already have osteoporosis, then take etidronate (Didronel). Current studies suggest it is effective. This drug has been used for years in the treatment of Paget's disease of the bone, a condition in which patchy areas of bone develop "holes" because their calcium has been absorbed. Didronel inhibits such reabsorption not only in patients with Paget's, but in postmenopausal women too, and it does so even after osteoporosis has begun. The result is increased bone thickness, especially in the spine, and fewer new fractures, all without significant side effects. I recommend a single 400 mg Didronel tablet a day, 2 hours before or after meals for 14 days every 3 to 3½ months—but for no longer than two years. It seems that prolonged use may be associated with an increase in fractures. You should also know that at the time of writing, the FDA has not yet approved Didronel for the treatment of osteoporosis. Most specialists, however, are prescribing it in the manner I have specified above. It's not cheap. Your druggist pays almost $150 for 60 such tablets. What he ends up charging you depends on your negotiating skills.

You can also decrease bone loss after menopause with calcitonin, a hormone that is made by cells in the thyroid gland. Unfortunately, it needs to be injected, so it's not practical for long-term therapy.

Diuretics, like hydrochlorothiazide (HydroDIURIL), commonly used for the treatment of high blood pressure and fluid retention, help prevent osteoporosis, too. However, I wouldn't take a diuretic with all its actual and potential side effects solely for that purpose.

Osteoporosis sets in long before the bones actually begin to fracture. There are special osteoporosis centers where scanning techniques are performed that will indicate how extensively your bones are involved even if you are still without symptoms. But these tests are expensive, and I don't see the point of having them unless you are at high risk. I think every woman should simply assume she is a candidate for osteoporosis later in life, and institute the measures I've listed above *before* the onset of menopause (lots of calcium in the diet, plenty of weight-bearing exercise), estrogen replacement *at menopause* (if there is no reason not to do so), and Didronel once osteoporosis has become apparent.

PALPITATIONS (ARRHYTHMIA): When You Ain't Got Rhythm

We are not normally aware of or concerned with our heartbeat as we go about our daily routine. But when the cardiac rhythm goes awry and out of synch, we experience palpitations (arrhythmia), which most individuals find alarming. Don't let the jumps, thumps, pauses, and pounding frighten you. They are usually harmless.

The normal heart beats regularly anywhere between 50 and 100 times per minute. In very fit individuals, athletes, and those who work out routinely and vigorously, as well as in persons taking certain medications that slow the heart, the rate may drop below 50 beats per minute. Conversely, when you're excited, frightened, have a fever, an overactive thyroid gland, anemia, or a variety of other illnesses (some cardiac, others not), the heart may beat faster than 100 times a minute.

Palpitations in the Absence of Heart Disease

Everyone has "extra heartbeats" now and then. Sometimes they are "silent" and you learn about them when your doctor examines you or records an ECG, but you perceive them as a thump in the chest or a pause in the heart rhythm. Such "premature" contractions can occur for no apparent reason, or they can be the result of anxiety, too much caffeine or thyroid supplements, appetite suppressants, or anti-asthmatic and a variety of other medications. It's a good idea to tell your doctor about them, although chances are he will not find that they are of any importance if you are in good health. In other words, if you *have been shown* not to have underlying heart disease, these symptoms are rarely a threat, and a doctor's reassurance is the only treatment most people need. But should these "harmless" palpitations continue to worry you, or leave you lightheaded or dizzy, there *are* drugs that will abolish them or reduce their frequency. Before deciding to resort to any of them, however, you should first eliminate caffeine, decongestants, and appetite suppressants (which you should never have started in the first place), have your thyroid dosage adjusted if you're taking any, discuss with your doctor the antidepression medication he has prescribed (some can induce or *worsen* rhythm abnormalities)—in fact, have *all* your medicines reviewed.

Do not rush into drug therapy to eradicate a harmless rhythm disturbance, because *every single anti-arrhythmic medication currently used can produce side effects,* some of which are potentially more dangerous and troublesome than the arrhythmia itself. So although premature contractions can almost always be abolished, it's often at a price.

If your arrhythmia consists not of an extra beat now and then, but a regular rapid pounding at a rate of 150, 160, 170, or more per minute, it needs to be stopped—something you may be able to do yourself. Take a deep breath, hold it, and then bear down as if you were moving your bowels. Or try gagging. Some doctors also advise pressing on the eyeballs or on the side of the neck. I think these latter maneuvers can be dangerous, and I don't encourage my patients to try them. *If the rapid beat continues for more than an hour, or if you become short of breath or have any chest discomfort, or break out into a cold sweat,* contact your doctor or go to a hospital emergency room straightaway. You may need an intravenous injection of one of the following: verapamil, adenosine, digitalis, or a beta-blocker drug—depending on the kind of rhythm disorder diagnosed in the electrocardiogram. Less commonly, an electric shock to the chest is required to terminate these attacks.

Once your rhythm has been "converted," you may be given an oral medication to prevent the arrhythmia from recurring. Which one, how much, and how long you should continue to take it depends on the specific kind of arrhythmia it was, how frequently you were experiencing it, how well you withstood it (did it make you short of breath or cause chest pain?), and whether or not you have underlying heart disease. If such maintenance treatment is needed, it's very likely to include digitalis. This drug comes in various forms and dosages, but the one I prefer is oral digoxin (Lanoxin). Most patients require a maintenance dose of 0.25 mg a day, sometimes more, sometimes less. You'll know you're taking too much if you gradually lose your appetite, have disturbances of vision, and—if you're male—your breasts enlarge and become tender.

There are several other oral anti-arrhythmic agents frequently prescribed for troublesome and recurring irregularities of heart rhythm. These include beta-blockers such as propranolol (Inderal), disopyramide (Norpace), quinidine (Quinaglute), procainamide (Pronestyl), and others. Such drugs must be administered at the discretion of your doctor because of their potential side effects. For example, if you're asthmatic, beta-blockers can make you wheeze; quinidine frequently causes diarrhea, hearing difficulties, or a rash; Norpace often interferes with urination in men with large prostates; Pronestyl can produce a serious form of arthritis.

Palpitations Accompanying Heart Disease

Even if you do have heart trouble, an arrhythmia should be treated only if it causes significant symptoms; *its mere presence does not usually warrant any therapy.* By "significant" symptoms, I mean shortness of breath, lightheadedness (when the brain isn't receiving enough blood because the heartbeat is too slow or too fast), or chest pain (because the cardiac rate is so rapid it tires the heart muscle, or so slow there's not enough blood being pumped to nourish it).

The choice of an anti-arrhythmic drug and its dosage may require consultation with a cardiologist experienced in this field because any drug that corrects heart rhythm may (a) not work in your particular case, (b) cause intolerable side effects, or (c) make things worse. Some medications even result in sudden "electrical" death because they can induce a "crazy" heart rhythm (a phenomenon called proarrhythmia).

There are several sophisticated techniques that can determine whether you actually need treatment for your arrhythmia and which agent is apt to be the safest and most effective for you. *The QRS signal averaging machine,* which looks very much like an ordinary electrocardiograph, may provide the answer in your doctor's office. If it doesn't, and the rhythm disturbance appears to be serious, you may require admission to the hospital for testing by a technique called *electrophysiologic stimulation (EPS)*. A catheter with an electrode at its tip is introduced through a vein in your arm or leg and threaded into the heart where it delivers a programmed electrical shock. This is calculated to bring on the arrhythmia, which the cardiologist treats right then and there with a variety of agents to see which one actually restores normal rhythm and/or prevents the electrical stimulus from provoking the arrhythmia again. Without such electrophysiologic stimulation, selecting the right drug is a matter of trial and error.

There's a fairly common arrhythmia called *atrial fibrillation* in which the heartbeat is totally irregular and usually quite rapid. It has become a household term since President Bush suffered his attack while jogging. Atrial fibrillation sometimes occurs in a completely normal heart, or it may be provoked by an overactive thyroid gland (as was the case with Mr. Bush), too many thyroid pills, other stimulants, and certain forms of heart disease. Regardless of its origin, you have several options if you develop atrial fibrillation. Your doctor may try to abolish it with drugs, or if that doesn't work, by electrical shock (cardioversion). If treatment is successful and your rhythm returns to normal, you may need to take something indefinitely to prevent recurrence—quinidine, disopyramide (Norpace), or procainamide (Pronestyl). But in some persons we simply cannot stop atrial fibrillation for very long no matter what drugs we use. You can safely live out the rest of your life with this irregular rhythm provided that the *rate* at which the heart is beating is not too rapid. That can usually be assured by taking digitalis and/or other drugs indefinitely. Since atrial fibrillation can lead to blood clots anywhere in the body, clots that cause a stroke if they reach the brain, *anticoagulants* (preferably warfarin, such as Coumadin) *or at least aspirin* are also required. Digitalis causes side effects—nausea, too slow a heart rate, and visual disturbances when taken in excess—that can be determined only by a blood test. Warfarin's main danger is bleeding when it makes the blood too thin. Monitoring this effect at intervals of 2 to 4 weeks is essential if you're taking this drug.

In summary, here are the key facts to remember. A cardiac rhythm disorder should not be treated just because it's there. If you can live with it and it's not the dangerous kind, you're better off leaving it alone. However, if it causes you symptoms, treatment may be necessary. Every drug used for this purpose has potential for harm, and the decision concerning which one to take requires sophisticated testing and consultation with a qualified specialist.

PANCREATIC CANCER: I Wish There Was a Treatment

The statistics for pancreatic cancer are dismal. It is the eighth most common malignancy in the United States; there are some 30,000 new cases a year, and there is no effective treatment. Only 2 percent of all patients live 5 years after the diagnosis has been made, and the majority die within 6 months.

The risk of cancer of the pancreas increases with age, so that while only 2 in every 100,000 people between ages forty and forty-five will contract it, it strikes 1 in 1,000 above fifty years of age. Some of the other statistics associated with pancreatic cancer are baffling. For example, why are men twice as vulnerable as women? Why do blacks have a greater and ever-increasing incidence than whites? What, exactly, is the relationship between pancreatic cancer and cigarette smoking? The connection is not nearly as strong as it is with lung cancer, but there does appear to be a slightly greater vulnerability among heavy smokers. Too much caffeine was once thought to be associated with pancreatic cancer, but the statistics do not support this link.

Recurrent or chronic pain in the belly without obvious cause should at least suggest the possibility of cancer of the pancreas, especially if it *penetrates through to the back*. In attempting to make the diagnosis, your doctor may not find anything "wrong" in the routine gastrointestinal X rays, but an abdominal sonogram or CT scan will usually confirm the diagnosis. Although by that time it is almost always too late for a cure, there is a faint glimmer of hope if the cancer has not yet spread. It all depends on its location within the pancreas. This organ, situated deep inside the abdomen, extends from left to right and consists of a "head" (on the right side), a "body" (the middle), and a "tail" (on the left). Occasionally a cancer localized to the *head* of the pancreas can be successfully removed; others cannot. It's a long shot, but worth exploring.

Radiation therapy sometimes prolongs life, but has never cured any case of pancreatic cancer of which I am aware. In some research centers, radiation-emitting pellets are being implanted into the cancer. This technique is still highly experimental, and the results are less than compelling. *Chemotherapy* may make you feel a little better for a while, but has very little impact on survival, and unfortunately, is not without side effects. The fact is, only *very* early detection of a cancer situated in a specific area of the pancreas permits even the faintest hope of a surgical cure. I pray that subsequent editions of this book will have happier news about this terrible malignancy. You can learn of any new breakthrough by calling the Cancer Hotline at 1-800-4-CANCER.

PARKINSON'S DISEASE: New Treatment, New Hope

Parkinson's disease is an affliction of the elderly and has four major components: a *tremor* at rest that disappears or lessens when you use your hands for some purpose like reaching for an object; *rigidity* of the arms and legs; *slowing* of all body movements; and a *loss of balance* with *difficulty walking*. It's easy to spot people with the full-blown condition: Their walk is really a shuffle, they're stooped forward, their face has virtually no expression, their voice is a monotone, and they often drool. But as in most other diseases, there is a spectrum of severity in Parkinson's ranging from symptoms that are very subtle, to those that are literally crippling.

Don't self-diagnose Parkinson's. You may end up living with a myth! If you suspect that that's what you have, see your doctor. And even if he confirms your fears, don't despair. Although Parkinson's is a progressive and debilitating condition, there are newer drugs that can improve both the quality and the duration of your life.

Parkinson's is due to a deficiency within the brain of a chemical called dopamine. Logic dictates that the best treatment would simply be to replace it (like insulin in diabetes, or thyroid hormone in hypothyroidism). Unfortunately, that's easier said than done. Many chemicals and drugs circulating in the blood that can reach every *other* part of the body are blocked from entering the brain by the "blood-brain barrier." Unfortunately, dopamine is on that "banned" list, and there is no way of getting it into the brain. But there is a chemical called L-dopa, which closely resembles dopamine, that *can* cross that barrier. L-dopa, available since 1967, has changed the face of treatment and improved the quality and length of life for patients with Parkinson's disease. It is a precursor (building block) of dopamine, and has many of its properties. Unfortunately, although the blood-brain barrier allows L-dopa to pass, very high doses are required to have any clinical benefit, and the amount of the active chemical needed to control the symptoms of the disease are toxic to the brain. But when L-dopa is combined with another drug called carbidopa, much lower doses will suffice. This combination of L-dopa and carbidopa (Sinemet) comes in various strengths (25 mg L-dopa and 100 mg carbidopa; 25 mg L-dopa and 250 carbidopa) and most patients require 3 doses daily. Certain dietary proteins (amino acids) like L-tryptophan, found in milk, and L-tyrosine, present in aged cheese and red wine, reduce the effectiveness of Sinemet, so you should avoid them if you're taking the drug.

There are two important drawbacks to Sinemet: (a) it has side effects, and (b) it becomes less effective with time. Nightmares, hallucinations, and other behavioral changes are the most troublesome adverse reactions of Sinemet. Also, if you've been taking the drug for a while and then stop, Parkinson's symptoms are greatly aggravated. But Sinemet's most discouraging property is its diminished effect after 2 to 5 years. As its potency begins to wane, the interval between doses becomes progressively shorter, so that unless you take the drug exactly on time, you may suddenly be immobilized, frozen, unable to move—something called the "on-off" phenomenon.

This 2- to 5-year limit in Sinemet's effectiveness has spawned two schools of thought with regard to its use. According to the first, since the duration of treatment is limited, the drug should be started as *late* as possible. Every other agent should be tried first until there is no alternative to the Sinemet. The other point of view holds that the 5-year limit is not the fault of the Sinemet itself, but a consequence of worsening disease. The conclusion here is that when Sinemet is started sooner rather than later, it slows down the progression of the disease and so continues to exert a beneficial effect for a longer time. Your doctor's point of view will determine the treatment you receive. It's important for you to discuss his stance in this matter in order to understand how your disease is being managed.

Regardless of when you start Sinemet, you'll be taking other drugs as well—before Sinemet therapy has begun, in combination with it, or after Sinemet is stopped. The most useful of these ancillary agents, in my opinion, is selegiline (Eldepryl) because it prolongs the effect of Sinemet, prevents its "breakdown" in the brain, and may also actually slow down the progression of the disease. The usual dosage is 5 mg, twice a day. It's the drug I recommend along with Sinemet when I first prescribe the latter. But never take Eldepryl with Demerol because of the possibility of serious side effects. The combination of Sinemet and Eldepryl makes it possible to use lower doses of the former drug at least initially, and so possibly delay the onset of the "wearing off" and "on-off" side effects.

Some years ago, researchers in the Soviet Union who were using an antiviral agent called *amantadine* (Symmetrel) for treatment of the flu noticed that it also improved the symptoms of Parkinson's disease. Further investigation revealed the fact that this drug stimulates the brain to make more of the missing dopamine! Amantadine is now widely used as a supplementary drug in treating Parkinson's.

A group of drugs called anticholinergics, the best known of which are Artane, Kemadrin, Cogentin, and Akineton, and certain antihistamines like Benadryl, can be added to the treatment regimen of some Parkinson's patients. These agents used to be the mainstay of therapy before L-dopa was discovered, but are not really widely used anymore. Their major action is on the tremor, but effective doses are accompanied by very troublesome side effects, including confusion, hallucinations, and retention of urine, particularly in the more elderly.

Now that you know some of the general principles of treatment, here's what I consider to be the best regimen for patients with Parkinson's disease. As soon as I make the diagnosis, I start my patients on amantadine, 100 mg, 3 times a day, and then wait for 2 to 3 weeks to see whether there is any improvement. If both the patient and I are impressed with the results, then this drug is continued for as long as it works. Do not take it if you have kidney disease, however, because a sick kidney can't excrete it and so the amantadine accumulates to undesirable levels in the blood. When amantadine's effect has begun to wane, I discontinue it and begin a combination of Sinemet, 25 to 100 mg, and Eldepryl, 5 mg, twice a day.

If these drug combinations still leave the patient miserable, I then try an antihistamine, like Benadryl or the anticholinergic Artane (but watch out for sleepiness, a very dry mouth, and difficulty voiding if you have a large prostate).

In addition to specific drug treatment, I advise my patients to take 3,000 units each of vitamin E and vitamin C a day. These are antioxidants that may, in such large amounts, retard the progression of symptoms and reduce the required dosage (and so the potential side effects) of most anti-Parkinson's drugs. There is an ongoing trial currently being conducted in twenty-six neurological centers in the United States, scheduled to be completed by 1995, which will either prove or disprove this theory. Until the final answer is in, I see no harm in taking these supplements. I also advise my Parkinson's patients to follow a low-protein diet, because it appears that eating more carbohydrates and less protein slows down the progress of the disease.

We may be on the verge of a new frontier in the treatment of Parkinson's disease with the latest experimental approach—transplantation of dopamine-producing fetal tissue directly into the brain of Parkinson's patients. In the last report I read, a Denver man so treated indicated that he was much improved after one year. This work is ongoing and worth monitoring.

PHLEBITIS: You've Got to Be Sure

Phlebitis refers to the inflammation of a vein and the resulting clot formation on its interior wall. You are at risk for developing phlebitis if:

- you're over age sixty;
- you have had major surgery recently;
- you are harboring a cancer of which you may or may not be aware (unexplained phlebitis may be the first evidence of a hidden, silent malignancy);
- you take oral contraceptives;
- you have varicose veins;
- you are obese;
- your blood flow is sluggish because your heart is weak (congestive heart failure);
- you have been lying around in bed for days without enough movement or physiotherapy (that's the reason doctors get you up so quickly after a heart attack or an operation);
- you've been on a long flight in one of those grotesque miniseats the airlines now provide, without periodically walking about;
- you have sustained a blow to your leg and have actually hurt the vein so that a clot forms within it.

Suspect phlebitis if your leg suddenly becomes painful, red, swollen, tender, and warm to touch, and the veins are visibly prominent. Given these classic symptoms, the diagnosis is obvious, and further testing may be unnecessary. But there are several other conditions that mimic phlebitis, so I almost always recommend an *ultrasound* test just to be sure because it's so simple and relatively inexpensive. The best one among the several techniques available, *color flow duplex sonography,* is about 85 percent accurate. If this procedure is not conclusive, you'll need a *venogram* in which a dye is injected into the vein—somewhat painful, and not entirely without a risk (though a very small one).

It is crucial to make the right diagnosis. You may not actually have phlebitis—for example, your leg may become warm, swollen, and tender after an injury but without clot formation in the vein, or you may have a local infection of the skin of the leg, again without vein involvement. On the other hand, the phlebitic vein may be situated so deep inside the thigh that although it hurts, there is nothing on the surface of the limb to indicate phlebitis—and so the diagnosis may be overlooked.

It's bad enough to call something phlebitis that isn't, but it's life-threatening to miss the diagnosis. A recent experience in my own practice highlights this point. A sixty-five-year-old man previously in good health called to say that he had struck his ankle with a piece of plywood. It was now red and throbbing, and so painful that he was unable to walk on it. Resting the leg, elevating it, and applying cold compresses afforded him only slight relief. He wondered whether the force of the blow might have chipped or cracked a bone, and asked me to arrange for X rays and orthopedic consultation at the emergency room of a hospital near his home. The specialist who examined him there found the ankle tender, swollen, and red, but there were no swollen veins visible, and an X ray did not reveal any bone injury or fracture. The doctor concluded that this was nothing more than a bad bruise. He prescribed an antibiotic, a pain killer, and sent the man home. The next evening the patient was found dead in bed—from a massive blood clot to the lung originating in veins near the "bruised" ankle! A proper diagnosis of phlebitis and immediate treatment might have saved this man's life.

Deep vein phlebitis requires anticoagulants. After you've been hospitalized, intravenous heparin, a fast-acting anticoagulant, is administered. It's important that the "thinness" of your blood be measured several times a day to determine the right dosage while receiving this drug. After 4 days of intravenous heparin, another anticoagulant, warfarin (Coumadin), taken by mouth, is added. The two agents are usually given together for another 4 days, or until the Coumadin effect is established, after which time the heparin is stopped and the Coumadin continued. It is absolutely essential that you have your blood checked at regular intervals (the test is called a "prothrombin time") while on this oral anticoagulant. Too little leaves you at risk for a traveling blood clot (embolism), while too much can cause internal bleeding. Also, get into the habit of always looking carefully for any evidence of bleeding in your urine, stool, and sputum, and report nose bleeds or unexpected bruise marks on the skin to your doctor. Anticoagulation, once begun, is usually continued for 3 to 6 months depending on the size and location of the involved vein, and whether there have been any complications, such as a piece of the clot breaking off and traveling to the lung (pulmonary embolism), where the venous circulation ends up.

Although a blood clot in the leg can be dissolved by one of the thrombolytic agents used to "melt" such obstructions in the coronary arteries during an acute heart attack (see page 119), I prefer anticoagulants for phlebitis because they are easier to use and are less risky. In any case, even if a thrombolytic agent were successful an anticoagulant would still be needed for follow-up to prevent the formation of new clots.

How does one treat phlebitis when anticoagulants can't be taken because of a bleeding ulcer or cancer, an abnormality of the blood-clotting mechanism, or recent surgery (when thinning of the blood can provoke a hemorrhage)? A small filter that looks like an umbrella can be easily inserted through one of the veins in your groin, and threaded into the large vein in the abdomen through which traveling clots pass on their way "north." This filter traps the emboli and prevents them from completing their journey to the lung. This procedure should also be considered if you've had blood clots to the lung despite adequate anticoagulation, something that does happen from time to time.

Not every case of phlebitis requires anticoagulants. If the inflamed vein is *very* superficially situated—that is, it's right there on the surface of the leg so that you can actually see and touch it—anticoagulants are not usually necessary. The best therapy for *superficial phlebitis* is indomethacin (Indocin)—an anti-inflammatory drug—50 mg, taken 3 times a day *after* meals. You needn't stay in bed, either. In fact, physical activity tends to dissolve such clots more quickly. I have my patients wear elastic stockings or bandages for support. If the inflammation does not subside within a couple of weeks (which it almost always does), I then refer them to a surgeon for possible removal of the affected vein. It's simple to do under local anesthesia, and does not require hospitalization.

In summary, phlebitis not only causes pain and swelling of the legs, it is also hazardous. The major danger is a traveling blood clot or embolus to the lungs. There are several new, accurate, and noninvasive techniques that can establish the diagnosis of phlebitis. As a rule, except when only the very superficial vessels are involved, patients require anticoagulants. Such drugs should be administered in a hospital, are not without risk, and once started must usually be continued for several months. The alternative is to dissolve a clot with thrombolytic agents, which, even if successful, still requires follow-up with anticoagulants. When it is not safe for you to receive a blood thinner, or if it has not been successful, a filter can easily be introduced into your large veins to trap any traveling blood clots before they reach the lungs.

PNEUMONIA: Treatment Depends on Which Kind You Have

Pneumonia is not a single disease but an infection or injury of the lung due to a variety of agents, infectious or chemical. These include scores of different bacteria, hundreds of viruses, several different fungi, and various parasites (like the protozoan that causes AIDS pneumonia), which are either inhaled or carried to the lungs in the bloodstream. Chemical injury is usually from toxic fumes and allergens. In addition, any foreign substances such as food, vomit, or liquid accidentally inhaled and entering the lungs—as sometimes happens when you swallow "the wrong way" or are in a drunken stupor, having an epileptic seizure, or unconscious—can cause "aspiration pneumonia." When you have pneumonia, regardless of the mechanism causing it, the lung itself is involved—in contrast to bronchitis, when only the air passages are infected.

Most pneumonias can now be cured by a wide variety of potent antibiotics. The best treatment depends on what is producing the pneumonia. The correct diagnosis requires a good history—given by you and taken by your doctor. For example, most infectious pneumonias caused by viruses or bacteria occur in the wintertime. But if yours came on during the summer, you may have been exposed to organisms in a moldy air conditioner and contracted Legionnaire's disease; if you work with or near birds, you may possibly have inhaled the psittacosis bug; if you've been traveling in the southwest U.S., perhaps you were infected by coccidioidomycosis, a fungus common in those parts; or maybe you were in the dusty central part of the country, the home of another fungus, histoplasmosis; if you're gay or use intravenous drugs, your pneumonia may be due to AIDS; if you live in a nursing home, you may have tuberculosis.

Pneumonia usually starts with a cough or fever, which may be low grade, like 101°, or high, in excess of 103°, and accompanied by shaking chills. In the first two or three days you may experience a sharp pain in the chest when you try to take a deep breath. That's *pleurisy,* an inflammation of the lining of the lungs covering the area of pneumonia, and it disappears as the pneumonia progresses. Soon, as you continue to cough, you will spit up infectious sputum, which is yellow, green, brown, or rust-colored (due to the presence of altered blood), or frankly bloody. You will also feel sick as hell.

That's the "typical" picture of pneumonia. Quite often, however, symptoms are so mild as to be mistaken for a bad cold—nothing more than a little fever, cough, and some sputum. It's only when the process continues for longer than you'd expect, or you begin to feel a little sicker than you should and decide to see your doctor, that the pneumonia is found. That scenario is called "walking pneumonia" because you've "walked" right through it.

In the real world of medicine, when you come down with pneumonia your doctor usually has a pretty good idea what kind it is, although he can't be absolutely sure until he completes the diagnostic process. In addition to a complete and thorough physical exam, you will need a chest X ray, sometimes a CAT scan of the lungs, various blood tests, and a sputum analysis to identify the offending organism. But while these data are pending, you should begin the therapy *most likely* to be effective in your case. In urban America, that's usually erythromycin, which is best for the pneumonia that follows the flu, or that due to Legionnaire's disease or another infectious agent called mycoplasma, and many other bacteria. However, after the bacteriology laboratory has actually identified the responsible infectious agent, your doctor may find that erythromycin was not the best choice and that a different antibiotic may be more effective—penicillin, one of the newer cephalosporins, tetracycline, ticarcillin (Timentin), etc.—the list is very long.

Pneumonia caused by an agent sensitive to one of the antibiotics rarely presents a problem. Treatment is usually continued for 7 to 10 days, or until you're without fever, have stopped coughing, and your X ray shows substantial improvement. Complete clearing of the film may not occur for several weeks. Aspiration pneumonia occurring as a consequence of coma due to poisoning, alcoholism, brain injury, and so forth, is more difficult to treat and takes longer. Here, the lungs are not only infected but are also irritated by whatever was inhaled.

Despite all the drugs available to cure infectious pneumonia, this is still a major killer, especially of the elderly. There is a vaccine that will reduce the risk of pneumonia caused by some 22 strains of pneumococcus, the most common infecting organism. I recommend it routinely for everyone over sixty years as well as for those at any age with chronic lung disease. It's a once-in-a-lifetime shot, and is remarkably free of side effects.

POISONING: Danger Lurks in Everyone's Home!

Only 15 percent of all cases of poisoning can be accounted for by suicide attempts. The remaining 85 percent are accidental—taking too much medication by mistake, or confusing one drug with another. Of course, there's also murder, for which I refer you to Agatha Christie, Mary Higgins Clark, or Stephen King. As you might expect, poisoning is most common in children under five, who will put anything that's available into their mouths, especially if it looks tempting and tastes good. At the other end of the spectrum, the elderly may consume the wrong drugs in the wrong amounts because they either can't see well or don't remember what they've already taken and when.

Substances most commonly overdosed are those you can buy without a prescription, like aspirin or Tylenol, household cleaning products, and pesticides. But then there are also sleeping pills, tranquilizers, "street drugs" like crack, and powerful legitimate prescription medication taken in error, to excess, or by the wrong person!

What to Do If You Witness a Poisoning, Or Are a Victim Yourself

Specific treatment for poisoning depends on what was consumed. In these pages I have not itemized antidotes for specific situations (you can and should get that information from your local poison control center), but you will find some important and practical pointers that will help you in a crisis.

Doctors, either in their offices or at home during the night, do not always have at their fingertips the information you need concerning antidotes when a poisoning has occurred. *So the very first thing to do is to contact the nearest poison emergency center.* You will find the one to call, wherever you live in the United States, in the list on pages 222 to 234. Keep that number by your telephone, especially if there are small children or elderly relatives living with you. If you live abroad, phone the local health department or medical society, police precinct, or hospital emergency room for the telephone number of a poison information office.

Administer *local treatment* as described in the following situations:

- If some chemical has gotten into the *eyes,* rinse them immediately with tap water for at least 20 or 30 minutes. If there's someone around to help you, have them hold the eyes open while you irrigate them. Begin this process even while the poison control center is being contacted.
- If the *skin* has come into contact with cyanide, a pesticide, or any chemical that may be absorbed or that burns or otherwise hurts it, *gently* rinse the area with water for 30 minutes, but not under pressure because that can force whatever is *on* the skin deeper inside.

If a poison has been *swallowed,* try to find out exactly what it was and how much was ingested. Then immediately *consult either your poison control center or your doctor* and ask whether it's safe to provoke vomiting. If it is, then do so immediately, even before symptoms develop. This will usually eliminate 50 percent of whatever is in the stomach. The best way to get someone to vomit is with a swig of syrup of ipecac—always keep some on hand. If ipecac is not available, put your finger or a spoon at the back of the throat. Remember how that makes you gag when your doctor examines you? *But never induce vomiting in the following circumstances:*

- If a caustic alkali or corrosive acid has been swallowed, because these substances damage the food pipe (esophagus) and stomach, which can be ruptured by retching and vomiting.
- In someone who is having convulsions (even if it is from the poison they took), who is unconscious, or in near-coma. Vomiting may get the poison into the lungs.
- In an infant under six months of age.
- In any child who has swallowed a toy, which may cause obstruction on its way out.
- In someone who has taken a quantity of sleeping pills and is already very drowsy. What they vomit can end up in the lungs.

Once you've administered local treatment to the eyes or skin, or have called your poison control center and induced the vomiting process if so advised, get yourself and/or the victim to the nearest hospital emergency room—regardless of the poison involved.

Poison Control Centers 1991

Centers in boldface have a 24-hour separate staff, and those that are also marked with an asterisk (*) meet the American Association of Poison Control Centers' criteria for certification. Centers in lightface lack those qualifications but handle a minimum of 2,000 information calls and 500 actual human exposures annually, except in areas where the population is too small to produce that number of calls. (Courtesy of *Emergency Medicine* magazine, Cahners Publishing Company, 1991.)

ALABAMA

Birmingham

The Children's Hospital of Alabama Regional Poison Control Center*
1600 Seventh Ave., S. 35233-1711
(800) 292-6678 (Alabama only)
(205) 933-4050
939-9201
939-9202

Tuscaloosa

Alabama Poison Control System, Inc.*
809 University Blvd., E. 35401
(800) 462-0800 (Alabama only)
(205) 345-0600

ALASKA

Anchorage

Anchorage Poison Center
Providence Hospital
3200 Providence Dr.
P.O. Box 196604 99519-6604
(800) 478-3193 (Alaska only)
(907) 261-3193

ARIZONA

Phoenix

Samaritan Regional Poison Center*
Good Samaritan Medical Center
1130 E. McDowell Rd., Ste. A-5
85006
(602) 253-3334

Tucson

Arizona Poison and Drug Information Center*
University of Arizona
Arizona Health Sciences Center
1501 N. Campbell Ave., Rm. 3204-K
85724
(800) 362-0101 (Arizona only)
(602) 626-6016

ARKANSAS

Little Rock

Arkansas Poison and Drug Information Center
University of Arkansas for
Medical Sciences
College of Pharmacy
Slot 522 (internal mailing)
4301 W. Markham St. 72205

(continued)

(501) 661-6161
(800) 482-8948 (MDs and hospitals) (Arkansas only)
(501) 666-5532 (MDs and hospitals)

CALIFORNIA

Fresno

Fresno Regional Poison Control Center*
Fresno Community Hospital and Medical Center
Fresno and R Sts., P.O. Box 1232 93715
(800) 346-5922 (Fresno, Kern, Kings, Madera, Mariposa, Merced, and Tulare counties only)
(209) 445-1222

Los Angeles

Los Angeles County Medical Association Regional Poison Control Center*
1925 Wilshire Blvd. 90057
(800) 777-6476
825-2722 (MDs and hospitals) (California only)
(213) 484-5151
664-2121 (MDs and hospitals)

Orange

UC Irvine Regional Poison Center
University of California, Irvine, Medical Center
101 The City Dr., Rte. 78 92668
(800) 544-4404 (Inyo, Mono, Orange, San Bernardino, and Riverside counties only)
(714) 634-5988
634-6665 (hazardous materials hotline)

Sacramento

UC Davis Medical Center Regional Poison Control Center*
2315 Stockton Blvd., Rm. 1511 95817
(800) 342-9293 (northern California only)
(916) 734-3692

San Diego

San Diego Regional Poison Center*
University of California,
San Diego, Medical Center
225 Dickinson St., H-925 92103-1990
(800) 876-4766 (Imperiale and San Diego counties only)
(619) 543-6000

San Francisco

San Francisco Bay Area Regional Poison Control Center*
San Francisco General Hospital
1001 Potrero Ave., Rm. 1E86 94110
(800) 523-2222 (415 and 707 area codes only)
(415) 476-6600

San Jose

Santa Clara Valley Medical Center Regional Poison Center
751 S. Bascom Ave. 95128
(800) 662-9886 (Monterey, San Benito, San Luis Obispo, Santa Clara, and Santa Cruz counties only)
(408) 299-5112

COLORADO

Denver

Rocky Mountain Poison and Drug Center*
645 Bannock St. 80204-4507
(800) 332-3073 (Colorado only)
525-5042 (Montana only)
442-2702 (Wyoming only)
(303) 629-1123

Interstate Centers

The Poison Control Center*
Omaha, Neb.
(800) 955-9119

CONNECTICUT

Farmington

Connecticut Poison Control Center
University of Connecticut
Health Center 06030
(800) 343-2722 (Connecticut only)
(203) 679-3473 (administration)
679-4346 (TDD)

DELAWARE

Wilmington

Poison Information Center
Medical Center of Delaware
Wilmington Hospital
501 W. 14 St. 19899
(302) 655-3389

DISTRICT OF COLUMBIA

Washington

National Capital Poison Center*
Georgetown University Hospital
3800 Reservoir Rd., N.W. 20007
(202) 625-3333
784-4660 (TTY)

FLORIDA

Jacksonville

St. Vincent's Medical Center
1800 Barrs St.
P.O. Box 2982 32203
(904) 387-7500
387-7499 (TTY)

Tallahassee

Tallahassee Memorial Regional Medical Center
1300 Miccosukee Rd. 32308
(904) 681-5411

Tampa

Florida Poison Information Center at Tampa General Hospital*
Davis Islands
P.O. Box 1289 33601
(800) 282-3171 (Florida only)
(813) 253-4444

GEORGIA

Atlanta

Georgia Regional Poison Control Center*
Grady Memorial Hospital
80 Butler St., S.E., Box 26066 30335-3801
(800) 282-5846 (Georgia only)
(404) 589-4400
525-3323 (TTY)

Macon

Regional Poison Control Center
Medical Center of Central Georgia
777 Hemlock St. 31208
(912) 744-1427
744-1146
744-1000

Savannah

Savannah Regional Poison Control Center
Department of Emergency Medicine
Memorial Medical Center
4700 Waters Ave. 31403
(912) 355-5228 (southeastern Georgia and southwestern South Carolina only)

Interstate Centers

Chattanooga Poison Control Center (for children only)
Chattanooga, Tenn.
(615) 778-6100 (northern Georgia only)

HAWAII

Honolulu

Hawaii Poison Center
Kapiolani Medical Center for Women and Children
1319 Punahou St. 96826
(800) 362-3585 (outer islands of Hawaii only) 362-3586
(808) 941-4411

IDAHO

Boise

Idaho Statewide Emergency Poison Communication Center
St. Alphonsus Regional Medical Center
1055 N. Curtis Rd. 83706
(800) 632-8000 (Idaho only)
(208) 378-2707

ILLINOIS

Chicago

Chicago & Northeastern Illinois Regional Poison Control Center
Rush-Presbyterian-St. Luke's Medical Center
1753 W. Congress Pkwy. 60612
(800) 942-5969 (northeastern Illinois only)
(312) 942-5969

Springfield

Central and Southern Illinois Regional Poison Resource Center
St. John's Hospital
800 E. Carpenter St. 62769
(800) 252-2022 (Illinois only)
(217) 753-3330

Interstate Centers

Cardinal Glennon Children's Hospital Regional Poison Center*
St. Louis, Mo.
(800) 366-8888 (western Illinois only)

INDIANA

Indianapolis

Indiana Poison Center*
Methodist Hospital of Indiana
1701 N. Senate Blvd. 46206
(800) 382-9097 (Indiana only)
(317) 929-2323
929-2336 (TTY)

Interstate Centers

Kentucky Regional Poison Center of Kosair Children's Hospital*
Louisville, Ky.
(502) 589-8222 (southern Indiana only)

IOWA

Des Moines

Variety Club Poison and Drug Information Center
Iowa Methodist Medical Center
1200 Pleasant St. 50309
(800) 362-2327 (Iowa only)
(515) 283-6254

Iowa City

Poison Control Center
University of Iowa Hospitals and Clinics
52242
(800) 272-6477 (Iowa only)
(319) 356-2922

Interstate Centers

McKennan Hospital Poison Center
Sioux Falls, S. Dak.
(800) 843-0505

KANSAS

Kansas City

Mid-America Poison Control Center
University of Kansas Medical Center
3900 Rainbow, Rm. B-400 66103
(800) 332-6633 (Kansas only)
(913) 588-6633

Wichita

HCA Wesley Medical Center
550 N. Hillside Ave. 67214
(316) 688-2277

Interstate Centers

Cardinal Glennon Children's Hospital Regional Poison Center*
St. Louis, Mo.
(800) 366-8888 (Topeka only)

KENTUCKY

Louisville

Kentucky Regional Poison Center of Kosair Children's Hospital*
P.O. Box 35070 40232-5070
(800) 722-5725 (Kentucky only)
(502) 589-8222 (metropolitan Louisville and southern Indiana only)

Interstate Centers

Knoxville Poison Control Center
Knoxville, Tenn.
(615) 544-9400 (southern Kentucky only)

MAINE

Portland

Maine Poison Control Center at Maine Medical Center
22 Bramhall St. 04102
(800) 442-6305 (Maine only)
(207) 871-2381 (ER)

MARYLAND

Baltimore

Maryland Poison Center*
University of Maryland School of Pharmacy
20 N. Pine St. 21201
(800) 492-2414 (Maryland only)
(301) 528-7701

MASSACHUSETTS

Boston

Massachusetts Poison Control System*
300 Longwood Ave. 02115
(800) 682-9211 (Massachusetts only)
(617) 232-2120

MICHIGAN

Detroit

Poison Control Center*
Children's Hospital of Michigan
3901 Beaubien Blvd. 48201
(800) 462-6642 (Michigan only)
(313) 745-5711

Grand Rapids

Blodgett Regional Poison Center*
Blodgett Memorial Medical Center
1840 Wealthy St., S.E. 49506
(800) 632-2727 (Michigan only)
356-3232 (TTY)
(616) 774-7854

Kalamazoo

Bronson Poison Center
Bronson Methodist Hospital
252 E. Lovell St. 49007
(800) 442-4112 (Michigan only)
(616) 341-6409

MINNESOTA

Minneapolis

Hennepin Regional Poison Center*
Hennepin County Medical Center
701 Park Ave. 55415
(612) 347-3141
337-7474 (TTY)

St. Paul

Minnesota Regional Poison Center*
St. Paul-Ramsey Medical Center
640 Jackson St. 55101
(800) 222-1222 (Minnesota only)
(612) 221-2113

Interstate Centers

McKennan Hospital Poison Center
Sioux Falls, S. Dak.
(800) 843-0505

North Dakota Poison Information Center
Fargo, N. Dak.
(701) 234-5575 (northwestern Minnesota only—call collect)

St. Luke's Midland Regional Medical Center
Poison Control Center
Aberdeen, S. Dak.
(800) 592-1889

MISSISSIPPI

Jackson

Regional Poison Control Center
University Medical Center
2500 N. State St. 39216
(601) 354-7660

MISSOURI

Kansas City

Children's Mercy Hospital
2401 Gillham Rd. 64108-9898
(816) 234-3000

St. Louis

Cardinal Glennon Children's Hospital Regional Poison Center*
1465 S. Grand Blvd. 63104
(800) 366-8888 (Missouri, western Illinois, and Topeka, Kan.)
392-9111 (Missouri only)
(314) 772-5200
577-5336 (TTY)

MONTANA

Interstate Centers

Rocky Mountain Poison Center*
Denver, Colo.
(800) 525-5042

NEBRASKA

Omaha

The Poison Control Center*
Childrens Memorial Hospital
8301 Dodge St. 68114
(800) 955-9119 (Nebraska only)
(402) 390-5555

Interstate Centers

McKennan Hospital Poison Center
Sioux Falls, S. Dak.
(800) 843-0505

NEW HAMPSHIRE

Hanover

New Hampshire Poison Information Center
Dartmouth/Hitchcock Medical Center
2 Maynard St. 03756
(800) 562-8236 (New Hampshire only)
(603) 646-5000

NEW JERSEY

Newark

New Jersey Poison Information and Education System*
Newark Beth Israel Medical Center
201 Lyons Ave. 07112
(800) 962-1253 (New Jersey only)
(201) 923-0764
926-8008 (TTY)

NEW MEXICO

Albuquerque

New Mexico Poison and Drug Information Center*
University of New Mexico
87131
(800) 432-6866 (New Mexico only)
(505) 843-2551

Interstate Centers

El Paso Poison Control Center
El Paso, Tex.
(915) 533-1244 (southern New Mexico only)

NEW YORK

Buffalo

Western New York Regional Poison Control Center at Children's Hospital of Buffalo
219 Bryant St. 14222
(800) 888-7655 (western New York only)
(716) 878-7654
878-7655

East Meadow

Long Island Regional Poison Control Center*
Nassau County Medical Center

(continued)

2201 Hempstead Tpk. 11554
(516) 542-2323
542-2324
542-2325
(911) (TTY)

New York

New York City Poison Control Center*
455 First Ave., Rm. 123 10016
(212) 340-4494
764-7667

Nyack

Hudson Valley Regional Poison Center
Nyack Hospital
N. Midland Ave. 10960
(800) 336-6997 (518 and 914 area codes only)
(914) 353-1000

Rochester

Life Line/Finger Lakes Regional Poison Control Center
University of Rochester Medical Center
Box 777 14642
(800) 333-0542 (Ontario and Wayne counties only)
(716) 275-5151
275-2700 (TTY)

Syracuse

Central New York Poison Control Center
University Hospital at Syracuse
750 E. Adams St. 13210
(800) 252-5655 (outside Onondaga County)
(315) 476-4766

Interstate Centers

Northwest Regional Poison Center
Erie, Pa.
(800) 822-3232 (southwestern New York only)

Vermont Poison Center
Burlington, Vt.
(802) 658-3456

NORTH CAROLINA

Asheville

Western N.C. Poison Control Center
Memorial Mission Hospital
509 Biltmore Ave. 28801
(800) 542-4225 (North Carolina only)
(704) 255-4490

Charlotte

Mercy Hospital Poison Control Center*
2001 Vail Ave. 28207
(704) 379-5827

Durham

Duke University Regional Poison Control Center
Duke University Medical Center
P.O. Box 3007 27710
(800) 672-1697 (North Carolina only)
(919) 684-8111

Greensboro

Moses H. Cone Memorial Hospital
Triad Poison Center
1200 N. Elm St. 27401-1020
(800) 722-2222 (Alamance, Forsyth, Guilford, Rockingham, and Randolph counties only)
(919) 379-4105

Hickory

Catawba Memorial Hospital
Poison Control Center
810 Fairgrove Church Rd., S.E. 28602
(704) 322-6649

Interstate Centers

Virginia Poison Center
Virginia Commonwealth University
Box 522, MCV Station
Richmond, Va. 23298
(800) 552-6337 (804 area code only)

NORTH DAKOTA

Fargo

North Dakota Poison Information Center
St. Luke's Hospitals
720 Fourth St., N. 58122
(800) 732-2200 (North Dakota only)
(701) 234-5575 (local; northwestern Minnesota only—call collect)

Interstate Centers

St. Luke's Midland Regional Medical Center
Poison Control Center
Aberdeen, S. Dak.
(800) 592-1889

OHIO

Akron

Akron Regional Poison Control Center
Children's Hospital Medical Center of Akron
281 Locust St. 44308
(800) 362-9922 (Ohio only)
(216) 379-8562
379-8446 (TTY)

Cincinnati

Regional Poison Control System and Drug & Poison Information Center*
University of Cincinnati College of Medicine
231 Bethesda Ave., M.L. 144 45267-0144
(800) 872-5111 (Ohio only)
(513) 558-5111

Cleveland

Greater Cleveland Poison Control Center
2101 Adelbert Rd. 44106
(216) 231-4455

Columbus

Central Ohio Poison Center*
Children's Hospital
700 Children's Dr. 43205
(800) 682-7625 (Ohio only)
(614) 228-1323
228-2272 (TTY)

Dayton

Western Ohio Poison and Drug Information Center
Children's Medical Center
1 Children's Plaza 45404-1815
(800) 762-0727 (Ohio only)
(513) 222-2227

Lorain

Poison Control Center of Lorain County
Lorain Community Hospital
3700 Kolbe Rd. 44053
(800) 821-8972 (Ohio only)
(216) 282-2220

Toledo

Poison Information Center of Northwest Ohio
Medical College of Ohio Hospital
3000 Arlington Ave. 43614
(419) 381-3897

Youngstown

Mahoning Valley Poison Center
St. Elizabeth Hospital Medical Center
1044 Belmont Ave. 44501
(800) 426-2348 (in Ohio: Ashtabula, Columbiana, Mahoning, and Trumbull counties only; in Pennsylvania: Mercer and Lawrence counties only)
(216) 746-2222
746-5510 (TTY)

Interstate Centers

Northwest Regional Poison Center
Erie, Pa.
(800) 822-3232 (northeastern Ohio only)

OKLAHOMA

Oklahoma City

Oklahoma Poison Control Center
Children's Hospital of Oklahoma
940 N.E. 13 St. 73104
(800) 522-4611 (Oklahoma only)
(405) 271-5454

OREGON

Portland

Oregon Poison Center*
Oregon Health Sciences University
3181 S.W. Sam Jackson Park Rd. 97201
(800) 452-7165 (Oregon only)
(503) 494-8968

PENNSYLVANIA

Allentown

Lehigh Valley Poison Center
Allentown Hospital
17 & Chew Sts. 18102
(215) 433-2311

Altoona

Keystone Region Poison Center
Mercy Hospital
2500 Seventh Ave. 16603
(814) 946-3711

Danville

Susquehanna Poison Center
Geisinger Medical Center
N. Academy Ave. 17821
(800) 352-7001 (Pennsylvania only)
(717) 275-6119

Erie

Northwest Regional Poison Center
Saint Vincent Health Center
232 W. 25 St. 16544
(800) 822-3232 (northwestern Pennsylvania, northeastern Ohio, and southwestern New York only)
(814) 452-3232

Hershey

Capital Area Poison Center
University Hospital
Milton S. Hershey Medical Center
University Dr. 17033
(800) 521-6110
(717) 531-6111
531-6039

Philadelphia

Delaware Valley Regional Poison Control Center*
1 Children's Ctr.
34th & Civic Center Blvd. 19104
(215) 386-2100

Pittsburgh

Pittsburgh Poison Center*
Children's Hospital of Pittsburgh
1 Children's Pl.
3705 Fifth Ave. at DeSoto St. 15213
(412) 681-6669

Interstate Centers

Mahoning Valley Poison Center
Youngstown, Ohio
(800) 426-2348 (Lawrence and Mercer counties only)

RHODE ISLAND

Providence

Rhode Island Poison Center*
Rhode Island Hospital
593 Eddy St. 02903
(401) 277-5727

SOUTH CAROLINA

Columbia

Palmetto Poison Center
University of South Carolina
College of Pharmacy 29208
(800) 922-1117 (South Carolina only)
(803) 765-7359

Interstate Centers

Savannah Regional Poison Control Center
Savannah, Ga.
(912) 355-5228 (southwestern South Carolina only)

SOUTH DAKOTA

Aberdeen

St. Luke's Midland Regional Medical Center
Poison Control Center
305 S. State St. 57401
(800) 592-1889 (South Dakota; also Minnesota, North Dakota, and Wyoming)
(605) 622-5678

Sioux Falls

McKennan Hospital Poison Center
800 E. 21 St.
P.O. Box 5045 57117-5045
(800) 952-0123 (South Dakota only)
843-0505 (Iowa, Minnesota, and Nebraska only)
(605) 336-3894

TENNESSEE

Knoxville

Knoxville Poison Control Center
University of Tennessee Memorial
Research Center and Hospital
1924 Alcoa Hwy. 37920
(615) 544-9400 (eastern Tennessee and southern Kentucky only)

Memphis

Southern Poison Center, Inc.
848 Adams Ave. 38103
(901) 528-6048
522-5985 (administration)

Nashville

Middle Tennessee Regional Poison/Clinical Toxicology Center
1161 21st Ave.
501 Oxford House 37232
(800) 288-9999 (mid-Tennessee only)
(615) 322-6435 (Nashville only)

TEXAS

Dallas

North Texas Poison Center*
Parkland Hospital
5201 Harry Hines Blvd.
P.O. Box 35926 75235
(800) 441-0040 (Texas only)
(214) 590-5000

El Paso

El Paso Poison Control Center
R.E. Thomason General Hospital
4815 Alameda Ave. 79905
(915) 533-1244 (southwestern Texas and southern New Mexico only)

Galveston

Texas State Poison Center*
University of Texas Medical Branch
Eighth & Mechanic Sts. 77550-2780
(800) 392-8548 (MDs and ambulance personnel only) (Texas only)
(409) 765-1420
(713) 654-1701 (Houston only)
(512) 478-4490 (Austin only)

UTAH

Salt Lake City

Intermountain Regional Poison Control Center*
50 N. Medical Dr., Bldg. 528 84132
(800) 456-7707 (Utah only)
(801) 581-2151

VERMONT

Burlington

Vermont Poison Center
Medical Center Hospital of Vermont
Colchester Ave. 05401
(802) 658-3456 (Vermont and bordering parts of New York only)
656-2721 (education programs)

VIRGINIA

Charlottesville

Blue Ridge Poison Center
University of Virginia Health Sciences Center
Blue Ridge Hospital
Box 67 22901
(800) 451-1428 (Virginia only)
(804) 924-5543

Richmond

Virginia Poison Center
Virginia Commonwealth University
Box 522, MCV Station
Richmond, Va. 23298
(800) 552-6337 (804 area code only)
(804) 786-9123

WASHINGTON

Seattle

Seattle Poison Center
Children's Hospital and Medical Center
4800 Sand Point Way, N.E.
P.O. Box C-5371 98105-0371
(800) 732-6985 (Washington only)
(206) 526-2121
526-2223 (TTY)

Spokane

Spokane Poison Center
St. Luke's Hospital
S. 711 Cowley 99202
(800) 572-5842 (northern Idaho, western Montana, and northwestern Oregon only)
(509) 747-1077

Tacoma

Mary Bridge Poison Center
Mary Bridge Children's Hospital
317 S. K St.
P.O. Box 5299 98405-0987
(800) 542-6319 (Washington only)
(206) 594-1414

Yakima

Central Washington Poison Center
Yakima Valley Memorial Hospital
2811 Tieton Dr. 98902
(800) 572-9176 (Washington only)
(509) 248-4400

WEST VIRGINIA

Charleston

West Virginia Poison Center*
West Virginia University
Health Sciences Center/Charleston Division
3110 MacCorkle Ave., S.E. 25304
(800) 642-3625 (West Virginia only)
(304) 348-4211

WISCONSIN

Green Bay

Green Bay Poison Control Center
St. Vincent Hospital
P.O. Box 13508 54307-3508
(414) 433-8100

Madison

University of Wisconsin Hospital Regional Poison Control Center
600 Highland Ave. 53792
(608) 262-3702

Milwaukee

The Poison Center
Children's Hospital of Wisconsin
9000 W. Wisconsin Ave.
P.O. Box 1997 53201
(414) 266-2222

WYOMING

Interstate Centers

Rocky Mountain Poison Center*
Denver, Colo.
(800) 442-2702

St. Luke's Midland Regional Medical Center
Poison Control Center
Aberdeen, S. Dak.
(800) 592-1889

For *Preventing* Accidental Poisoning:

- Put safety latches on every cupboard in which you keep *any* substance other than food. (If there are some really fat people at home, such a latch on the refrigerator isn't a bad idea either!)
- Identify *every* product in your house very clearly. Don't fill empty containers with materials that don't belong in them. For example, never put cleaning fluid in a soft drink bottle. Someone mistakenly drinking the contents might die. When an original container is empty, throw it out, recycle it, or return it for the deposit, but don't use it again.
- Never coax children to take medicine by telling them it's candy or delicious. You may be so convincing that when you're away from home they will look for and find whatever it is you've encouraged them to take, and finish it off.
- When a pill is due to be taken during the night, don't grope for it in the dark. Turn the light on and identify it properly even if it means waking whoever is sleeping next to you. Also, don't keep an array of drugs on your night table. You never know what you might swallow inadvertently when you're sleepy.
- Unless you have severe arthritis affecting your hands, or bad Parkinson's with a tremor, or are paralyzed from a stroke, ask your druggist to dispense all your medications, whether prescription or over-the-counter, in child-resistant containers with caps.
- Don't leave a colorful array of detergents, drain cleaners, and other such products lying around your kitchen, the laundry area, the bathroom, or the garage. I recently looked into my own garage to see whether I myself was guilty of the very things I'm advising you not to do. To my horror, I found some cans of paint and rust remover, insect sprays, weed killers, fertilizers, antifreeze, and even a spare can of gasoline, all lying within easy reach. Even though my children are grown and we don't yet have any grandchildren, what a disaster it could have been if some neighbor's kid had wandered into my garage and decided to taste what I'd left around! In any event, I've locked every one of those items away in storage bins and closets, out of the reach of any inquisitive child. While on the subject of garages, remember that many children are pretty precocious these days, as exemplified by the ten-year-old who flew an airplane from coast to coast not long ago. Kids love to climb into cars and turn a key that's been left in the ignition. If the garage door happens to be closed, that child could be

poisoned by carbon monoxide. So always remove the keys from your car, minibike, snow blower, or any other gasoline-powered machine.

- Windowsills are no place for medications. Any item that requires such "cold storage" should be placed in a child-resistant container at the *back* of your refrigerator.
- When you review the status of your medicine cabinet from time to time, don't throw the outdated drugs into a wastebasket where they can later be retrieved by your own children, or in a garbage can outside your house where the neighbors' toddlers can get at them. Flush them down the toilet.

POTASSIUM DEFICIENCY: Ugh!—Or Mmm?

Danny Kaye, the late great comedian, actor, and (anyone who knew him well would add) "philosopher," loved to play with words. *Potassium* was one of his favorites because it solved all kinds of social problems for him. For example, he would tell a police officer who had stopped him for speeding that he was on a "potassium emergency"; he'd leave a boring evening early in order to attend to his "potassium situation." A story I recounted in one of my earlier books bears repeating here. Danny (who at the time was trying to lose weight) and I (who should have been) were dining at the home of a mutual friend. Our hostess was a superb, generous, and enthusiastic chef, and this was not the time or place to be counting calories. After the meal itself, which was both irresistible and plentiful, the coup de grace came in the form of an enormous banana cream pie presented with great pride and fanfare. Our hostess, who let it be known in no uncertain terms that she had baked it herself in honor of Danny's visit, was very disappointed when he demurred. She: "It's my special creation!" Danny: "I just can't handle it." She: "Please, just taste it." Danny: "I'm sorry, I don't think I can." She: "You don't know how you're disappointing me." At this point, Danny leaned over and whispered very confidentially in the good lady's ear, "I'd hoped not to have to tell you this, but you've left me no choice. I have a potassium problem!" She was shocked. "Oh, no! You do? Well, then of course I understand." She removed the hefty slice of banana cream pie, and never said another word about it. But I, without any such alibi, was expected to eat my share—all of it! I loved it!

Potassium supplements are the "in" thing right now, and I think more people are taking them than need to. If you're really deficient in potassium because of prolonged diarrhea or vomiting; if you've got the specific kind of kidney disease in which too much potassium is excreted in the urine; or, if you've been taking diuretics, which lower the potassium level in the blood, you may require additional amounts. You will know you do if you're weak, tired, experience numbness and tingling in your hands and feet, or develop palpitations. Normally, you can make up for the loss by eating more potassium-rich foods, like meat, fish, bananas, prunes, raisins, apricots, spinach, potatoes, and beans. But if your doctor tells you that that's not enough, and that you need a supplement, the best among the many available is *potassium chloride* (because when the body loses potassium, it takes chloride with it, so you're apt to be deficient in that, too). None of the other commercial potassium supplements—gluconate, citrate, bicarbonate, or acetate—contain chloride. Unfortunately, potassium chloride tastes awful in liquid form. But if your druggist is imaginative, talented, and takes the time, he can concoct a preparation that at least won't make you gag. If he won't bother, there are two readymade preparations you can buy that cost a little more but almost border on the tasty—Elixir Kaon and K-Lyte-C1. Take them *after* meals (because potassium is a gastric irritant) in whatever dose is prescribed, and feel free to dilute them as long as you empty the glass.

Some of my patients just can't stand liquid potassium, and insist on a tablet or capsule. If you too prefer that, take the slower timed-release preparations. They're less irritating to the stomach. But once you have swallowed such a capsule, avoid all antispasmodics, which slow down the intestinal waves that propel the gastric contents along. Unless you do, the slow-release medication may dissolve in the stomach and erode its lining. Also, make sure before you take any potassium supplement that your diuretic doesn't already contain either potassium or a potassium-sparing substance. Many of them, like Dyazide, Maxzide, Aldactazide, and Moduretic, do. You don't want to end up getting more potassium than is good for you. Certain blood pressure–lowering pills such as the ACE inhibitors (Capoten, Vasotec, Prinivil, Zestril), though not diuretics, also tend to raise the potassium level and should not be supplemented either.

PREMENSTRUAL SYNDROME: When You're *Not* Likely to Make Friends and Influence People

Nine of every ten women sometime or other during their reproductive years experience unpleasant symptoms a few days before their period begins. Although such premenstrual symptoms (PMS) cover the spectrum of human complaints, they are fairly constant in any given female. I recently came across a roster of 150 different symptoms, the most common of which were depression, anxiety, unprovoked hostility and anger, abdominal pain, cramps, weight gain, acne, craving for sweets (especially chocolate and salty foods), irritability, headache—and on and on. Miraculously, they all disappear with the onset of the period.

No one has identified the actual mechanisms responsible for the premenstrual syndrome, but you can be sure it's hormonal and not "in the head." Even the courts recently agreed and acquitted a woman accused of murder on the grounds that she was suffering from premenstrual syndrome at the time she committed the crime!

Hundreds of remedies have been proposed for the relief of premenstrual symptoms over the years. Those still touted today by many women and some doctors include megadoses of vitamin B_6 (pyridoxine), evening primrose oil, vitamin supplements, dietary manipulation, extra magnesium, progesterone, oral contraceptives, synthetic hormones, bromocriptine (Parlodel), danazol (Danocrine—used in the treatment of endometriosis), spironolactone (Aldactone—a diuretic), and even the antibiotic doxycycline (Vibramycin). Among these, currently the most popular is probably vitamin B_6. There's no theoretical reason why it should work, and *I do not recommend it* because it can hurt the nervous system, even in doses smaller than the 2,000 mg a day "recommended" for the treatment of PMS. Evening primrose oil, another favorite, has not passed any scientific evaluation either. If you're determined to try it, don't do so on an empty stomach because it can cause gastric irritation. High doses of magnesium will give you diarrhea; megavitamins can be toxic, too.

There *are* several approaches to PMS that do, at least, have some theoretical rationale, if no proven efficacy, and are worth trying:

- Spironolactone (Aldactone), a mild diuretic, 25 mg twice a day, may remove some of the fluid retained at this time of the cycle.

- Danazol (Danocrine) decreases estrogen levels and may alleviate PMS. Take 100 to 400 mg, twice daily for 3 to 9 months.
- Doxycycline (Vibramycin), an antibiotic, was found to benefit 50 percent of women in one study (possibly by eliminating low-grade uterine infection). The dose is 100 mg a day.
- Vitamin A, 5,000 units, 3 times a day, has diuretic and anti-estrogen properties, and is recommended by some gynecologists.
- Progesterone suppositories, 10 mg daily for 1 to 2 months (taken orally, natural progesterone is destroyed in the gastrointestinal tract), is a favorite of some doctors, but I've never seen it work.
- Bromocriptine (Parlodel), a 2½ mg tablet twice a day for 10 to 14 days before menstruation, may help by virtue of its hormonal action.
- Dietary changes—the avoidance of alcohol, caffeine (lots of agreement on this), sugar, and salt and a decrease in fat intake—do seem to help. Whole grains, fresh fruits and vegetables, lots of seeds and nuts, combined with a regular exercise program are also beneficial.

But in the final analysis, the one sure thing to cure PMS is menopause!

PROSTATE CANCER: Every Man's Disease

Your chances of developing cancer of the prostate gland are about one in eleven if you're white, and one in seven if you're black. The incidence of death from this malignancy is second only to lung cancer in the United States, and about 30,000 men die from it annually (over 100,000 new cases are diagnosed each year). Those who experience symptoms represent only the tip of the iceberg. Most cases of prostate cancer are silent, and are diagnosed only incidentally after a gland that was removed, not because it was cancerous but because it was large and interfered with normal urination, is looked at under the microscope and found to contain malignant cells. These cells are localized within the prostate, and may never have gone on to cause any trouble even if they hadn't been discovered. But, in other instances in the course of a routine rectal examination, your doctor may feel a lump with his finger that turns out to be malignant when biopsied. Unfortunately, by that time the cancer may already have spread to other parts of the body.

My professor of surgery used to tell us, "No physical exam is complete without a rectal. If you don't put your finger in, you put your foot in." In the course of such an exam, your doctor feels for two things—cancer of the rectum and anus, and more important, cancer of the prostate.

Most individuals don't enjoy having a rectal exam, and some of my own patients try to negotiate me out of doing it. "What's the point, doctor? I hardly ever have to get up at night to empty my bladder except when I've had some beer. I can still "pee" over a fence, and my sexual function is great. Why don't you just leave me alone?" I tell them, as I am performing the examination (which I always do despite the plea bargaining), that the need to get up more than once a night, frequent urination, a puny stream, and recurrent infections of the prostate are all signs of an *enlarged* gland—an entirely different kettle of fish from cancer. *Prostate cancer is usually silent until late in the game,* because it doesn't often enlarge the whole gland itself and so the symptoms of urinary blockage are absent.

If your doctor finds a suspicious lump, he will draw some blood to measure the level of *prostatic specific antigen* (PSA), an indicator of malignancy. Because prostate cancers are sometimes too small to be felt, I have been performing the PSA test routinely on all my male patients over age fifty-five for some three years now. It has turned up silent prostate malignancy in several men in whom there was no other evidence—much as a mammogram picks up tiny breast cancers. The *acid phosphatase test,* which tells us whether or not the prostate cancer has already spread to bone, is done if the PSA level is abnormally high. Even if both these assays are normal, the lump may still be malignant. Although it may also reflect a prostate stone or scar tissue from an old infection, one can't be sure, which is why it is usually necessary to have tissue confirmation. This can be obtained in one of two ways. The first is an office procedure in which the contents of the nodule are sucked out with a very fine needle (fine needle aspiration—FNA). If the cells are identified as malignant *without question,* then the diagnosis is secure. But if the technique does not hit "pay dirt," the possibility of cancer is not excluded. You now need an actual biopsy in which a slice of the suspicious area is removed and studied. This can often be done in the urologist's office utilizing ultrasound guidance and the "biopsy gun"—two instruments with great diagnostic accuracy—or in the hospital itself.

If the diagnosis of prostate cancer is confirmed, you need a very thorough general exam, a chest X ray, as well as a bone scan to see whether the malignancy has spread beyond the gland. The best treatment at that point depends on whether or not it has done so.

Let's consider the possibilities. If the cancer is still limited to the interior of the prostate, you have three choices.

• Option number one is to *leave it alone,* especially if you're old, a poor surgical risk, or if the cancer cells are well differentiated—that is, they don't appear to be "wild" or look as if they're about to take off. The aggressiveness of most tumors can be judged by how much they vary in appearance from the normal tissue from which they originate. The more they resemble such normal cells, the more slowly they're likely to spread. If you decide to wait, you should be re-evaluated every three or four months with a rectal examination (to see whether the lump has gotten any bigger), and a PSA blood test, which is a more sensitive indicator of cancer growth. If there is measurable progression, then further treatment is necessary. How risky is the wait-and-see course? In the most reliably reported studies to date, every cancer that was not treated as soon as it was discovered did eventually progress, especially in older men. But if you are over seventy-five years of age and/or have serious heart trouble, kidney disease, diabetes, or some other complication that makes treatment hazardous, time may actually be on your side and an aggressive approach is probably undesirable.

• If you decide to act as soon as the prostate cancer is found, Option number two is *radiation.* In my experience, for most men, X-ray treatment at this early stage offers a better long-term outlook than does "watchful waiting." As many as 85 percent of patients live at least ten more years. Why don't urologists advise radiation to everyone? Because even when performed by a very experienced radiologist, this treatment can cause troublesome side effects—chronic diarrhea, rectal pain, fissures, incontinence, infections, and impotence. In my view, none of these possibilities excuse delay in therapy.

• Option number three is the one that offers about the same chance for cure as radiation, maybe even a little more—*removal of the prostate by surgery.* Unfortunately, it's a serious operation that carries with it some risk, especially if you have some other serious underlying disease.

Which of the three options do I recommend? If you're under seventy and in good general health, I advise surgery. In all other instances, my preference is for radiation *by an expert.*

What's the best treatment *after the cancer has already spread?* None of the three options mentioned above are now likely to result in a cure. It's like locking the barn door after the horses have gone. What you must do now is reduce as much as possible the amount of male hormone (testosterone) being made by your body, because the cancer feeds on that hormone. You can prevent testosterone production, which occurs mainly in the testes, by removing them surgically, or you can take antitestosterone drugs. Surgical castration sounds brutal, but it's probably the better alternative for most men. The procedure itself is a very simple one. You may leave the hospital late the same day or the next, and for cosmetic purposes, small plastic balls are inserted into your scrotal sac so that no one in the locker room will ever know! Of course, that operation leaves you impotent, but so do nonsurgical alternatives. Either course will slow the rate of cancer growth and relieve pain, but it cannot cure you. However, don't minimize the importance of the slowing effect. One of my patients with cancer of the prostate that had already spread was diagnosed at age seventy-five and lived to be ninety-four. The only thing we ever did for him was to remove the testes.

If, in the course of your disease, you begin to experience pain in the bones involved by the cancer, radiation to the area provides substantial relief.

There is very active ongoing research in the area of prostate cancer. Always ask your urologist for the latest breakthroughs in medical or surgical approaches to this problem. Also, remember the Cancer Hotline: 1-800-4-CANCER.

PROSTATE ENLARGEMENT: When You Void the Way Some People Stutter

The following scenario is familiar to millions of men in their early sixties and older. Your previously undisturbed sleep has recently become punctuated by insistent calls to empty your bladder—two, three, or more times a night. At first you think that it may be because you've been drinking too much during the evening, but cutting back makes little difference. It also takes you longer to start voiding, and when you do, the stream is weak, sometimes split (which can be embarrassing when you wet someone else's trousers in a men's room), and is almost always followed by a little dribble after you think you are finished.

This constellation of symptoms almost certainly reflects enlargement of the prostate, something that happens in varying degrees to virtually every man as he gets older.

A third of all males in the United States, by the time they're eighty-five (some 400,000 strong), choose to have their big prostate gland surgically removed. But remember the days when any child whose tonsils were enlarged was sent for the ritual tonsillectomy? That's rarely done anymore because there are now better ways to deal with recurrent sore throats. The same surgical reticence does not yet apply to an enlarged prostate, but we're getting there. I predict that in the not too distant future, the prostatectomy will go the way of the tonsillectomy. Until then, what is the best treatment when you're up all night voiding?

Men with symptoms of prostatism currently have the following treatment choices: (a) surgery, (b) medication, or (c) one of the newer experimental nonsurgical techniques that are already being fairly widely performed. Here is what you should know about each of them before you decide what to do.

The Pros and Cons of Surgery

Prostatectomy, removal of prostate tissue, does relieve most of the symptoms of an enlarged prostate, and there are two different methods of doing it. The first, called a *transurethral resection* of the prostate (TURP), is a "shaving" or reaming of the gland. An instrument with a tiny light and knife at its end is inserted into the penis, and a portion of the overgrown prostate that is blocking the urinary passage is scraped away.

Some investigators are cutting the enlarged prostates away with lasers, but that's still surgery as far as I'm concerned, except that light energy is substituted for the knife. It carries the same risks and side effects as any other operative techniques. In the other method, the "*suprapubic*" or "open" procedure, an incision is made in the lower abdominal wall and the obstructing tissue is removed. Both operations leave some of the gland behind. Because the "open" surgery is more substantial than a TURP and carries with it a longer recuperation, most men choose the TURP. Also, the incidence of impotence may be higher after the suprapubic operation. That may not be too important for most men in their eighties (although I know some octogenarians who would vehemently dispute that statement), but it can be disastrous to a man in his fifties or sixties.

A down side to both surgical approaches is the occasional failure to relieve the symptoms of prostatism to any significant degree—some men still have to "go" frequently during the day and night, and a small number become incontinent because the muscles that control the flow of urine were weakened or damaged during the surgery. Occasionally the procedure causes the formation of scar tissue in the urinary duct within the penis, interfering with the ability to void freely. Also, during ejaculation, the sperm may travel back into the bladder instead of forward out of the penis because of damage to the muscles in the area, but this "dry event" does not appreciably decrease the pleasure of orgasm. One man in five who has undergone the TURP procedure needs another operation within 5 years because some of the prostate tissue grows back. Moreover, although the TURP does reduce the symptoms of an enlarged prostate, it does not prolong life. In fact, the only statistical study on this subject of which I am aware suggests that it actually reduces life span by one year. Neither type of operation eliminates the risk of developing cancer later on because a portion of the prostate gland left behind can become malignant.

In practical terms, this all means that if your doctor has advised you to have your prostate out, remember to ask if your symptoms are actually a threat to you, and whether it's safe to live with them. For example, if after you *think* you've emptied your bladder there are still several ounces of urine left behind, you are at risk for "acute urinary retention." That means that one day while you are on a train, a plane, climbing a mountain, driving on some remote highway, or scuba diving off a tropical island hours away from medical help, you may suddenly have to "go" and find that you can't! The signal to void becomes progressively more intense and painful, until you have to be rushed to the nearest medical facility where a catheter inserted into the bladder via the penis is the only way to release the retained urine. If help is a long way off, you may be in real trouble. So if you're vulnerable to such acute obstruction, then the option of trying medication to control your symptoms may be risky. Don't chance the sudden shut-down of your voiding mechanisms; opt for surgery.

There are other situations that mandate surgical intervention. If the back-up of urine from the enlarged prostate has affected your kidneys and is distending or blowing them up, you must correct matters quickly, before permanent damage results. Or, if you have repeated urinary infections, you're best rid of the obstructing prostate gland. However, there is no rush for surgery simply because your symptoms are inconvenient. I have many, many patients who continue to live with, and have adjusted to, an enlarged prostate, secure in the knowledge that their lives are not in danger. And the symptoms it causes can often be improved by medication.

Drug Therapy

If you decide to postpone an operation, ask your urologist about the following drugs: Phenoxybenzamine (Dibenzyline)—a 10 mg tablet at bedtime may significantly decrease the frequency of your visits to the john. Also effective is prazosin (Minipress), widely used in the treatment of high blood pressure, 1 mg at bedtime and then 1 or 2 mg more during the day. Before you commit yourself to this particular drug, however, always try a test dose of 1 mg at bedtime to see whether it drops your pressure too low. You will know it's not for you if you feel lightheaded and dizzy when standing.

Finally, there's an exciting new development, an agent called *Proscar,* now in active research by the Merck Company, that actually shrinks the enlarged gland in as many as 30 percent of cases (Dibenzyline and Minipress control symptoms, but do not reduce the size of the prostate). This experimental medication has a fascinating history. Some years ago it was observed that certain natives in Santo Domingo rarely if ever had prostate enlargement. Further study revealed them to be deficient in a specific enzyme normally present in other populations. This led some scientists to speculate that if they could come up with a drug that neutralizes this enzyme, we would all be like those Santo Domingo Indians, because our prostates would then either not increase in size, or would shrink if they had. Believe it or not, these researchers have succeeded in synthezing this substance and it does exactly what they hoped it would. In clinical trials so far, when given to men with prostatism, it appears that at least one in three may be spared surgery because the drug shrinks the big gland. Proscar is under continuing investigation to make sure that it is safe over the long term, but several urologists who have participated in the clinical trials have told me they are hopeful that it will pass the required tests. So if you have symptoms of prostatism, and seem to be headed for surgery, double check with your urologist about the status of this new Merck drug.

Experimental Nonsurgical Techniques

If surgery seems imminent and unavoidable for any reason (other than a real emergency), hold on! There are two interesting new alternatives to consider and discuss with your urologist. In the first, prostatic *ballooning*, a deflated cylindrical balloon at the end of a catheter, is inserted into the penis under local anesthesia. When it reaches the area of the prostate it is inflated for 10 to 15 minutes, compressing or pushing aside the enlarged prostatic tissue that's interfering with the outflow of urine—a procedure not unlike angioplasty of obstructed arteries in the heart and legs. Thousands of prostatic balloonings have already been successfully performed. Although it's still too early to tell how long this compression will hold, the technique appears to be simpler, safer, and cheaper than an operation, and most patients can go home the very next day.

Finally, applying *heat* to the enlarged prostate by means of a catheter introduced into the urethra of the penis may shrink it too, especially when done in combination with ballooning. These procedures are still "experimental" but available, and worth keeping in mind.

In summary, here's the bottom line. Before submitting to surgery for an enlarged prostate, ask whether it's absolutely necessary. If it's safe to wait, then do so. If something *must* be done, consider first one of the nonsurgical alternatives already available.

PSORIASIS: Relief, But No Cure

Three out of every hundred Americans suffer from psoriasis, a skin condition consisting of silvery-white flaky scales that can be found on virtually every part of the body—from the belly button to the scalp. These plaques may or may not itch, and are sometimes associated with a form of arthritis. For the most part, their main symptom is their cosmetic embarrassment. We don't know its cause, and there is no cure, but there are steps you can take that will help. Several of my patients have gone to the Dead Sea area of Israel and returned completely free of the scales that once plagued them. Three months later the lesions returned. I have no wish to hurt Israel's tourist trade, but the fact is that the same sunshine in which these individuals basked in the Holy Land can be delivered by a machine at home at a fraction of the cost. Before embarking on any therapy, however, make sure you are not on lithium for a manic-depressive state, or on a beta-blocker such as Inderal, both of which can aggravate the condition.

The best initial treatment for psoriasis is a topical steroid cream or ointment. The cream is more convenient during the day, but you should use the ointment at night because it is less drying. Start with the weakest strength (like Cortaid, which you can buy over the counter), and then cover the affected areas with plastic wrap. As time goes by, you will probably need progressively stronger preparations. After Cortaid has lost its effect, I prescribe fluocinonide (Lidex) or similar higher potency steroid creams and ointments. For more severe cases, Diprolene ointment often helps considerably. But don't use it for more than 2 weeks at a time because it is absorbed into the body to some extent, and can result in the usual adverse side effects of topical cortisone (see page 165). When one topical steroid preparation loses its potency, switch to another brand, but always check the strength of the new one you're trying. And incidentally, never apply any of these creams more than very occasionally to areas of the body where the skin is thin, such as the groin, the back of the hands, or the face, because prolonged use will damage the skin.

I've always found it helpful to use a tar preparation along with the steroid because should you ever need to stop the latter abruptly, the tar minimizes the severity of the "rebound phenomenon" (a worsening of any disorder when therapy is withdrawn). The tar preparation I prefer comes readymade in an over-the-counter cream called Fototar, but don't apply it to the armpit or the groin, where it irritates the skin. For psoriasis affecting the scalp, as it so often does, Selsun shampoo alternating daily with a tar shampoo like T-gel is effective. You can pick the Selsun shampoo right off the shelf, but Selsun lotion requires a prescription! Such are the ways of the FDA!

When psoriasis is widespread, you can apply these preparations to areas that are especially hard-hit, but it's not feasible to cover the entire body with topical steroids and tar. You now need a generalized treatment. There is a very old therapy called the Goeckerman technique, developed at the Mayo Clinic in the early 1900s, in which a combination of tar and certain light rays are used. This treatment will keep the psoriasis at bay for as long as 8 months, but it's a hell of a nuisance and a full-time job. You need either to be admitted to the hospital or to come back as an out-patient *every* morning and night for weeks.

If you're not up to making that kind of commitment, there is the alternative of UVA photochemotherapy for severe cases. First you take oral psoralen, which photosensitizes the skin, and then sit in a box that delivers ultraviolet A rays to the body. An important drawback of this therapy is the increased risk of skin cancer and cataracts with which it is associated. I'm not keen on it.

Psoriasis can be such a miserable business that almost any treatment with some promise should at least be considered. Sulfasalazine (Azulfidine), which is helpful in inflammatory bowel disease and rheumatoid arthritis, also appears to improve psoriasis. It has been reported to result in impressive clearing of the lesions in at least 50 percent of cases. It's all relatively new, and worth discussing with your dermatologist.

There is one additional recent development in the management of psoriasis of which you should be aware. Apparently, in severe cases there is an overproduction of human growth hormone (HGH) by the body. A product called Sandostatin has been shown to decrease the blood levels of HGH, and apparently improve psoriasis. These observations have thus far been made in "uncontrolled" trials, and are currently being tested more scientifically. If your psoriasis is severe, ask your doctor about the status of this Sandostatin research.

RAYNAUD'S DISEASE: Warm Body, Cold Fingers and Toes

There are some people, especially younger women, whose fingers, toes, and tip of the nose hurt when the weather is even just a bit cool. Also, just dipping the hands or feet into cool water makes them hurt, then tingle, and finally leaves them numb. They become discolored, too—first white, later blue. Repeated exposure to cold eventually leaves the tips of the fingers and toes permanently "wrinkled." While their extremities are affected in this way by a drop in temperature, the rest of the body remains comfortably warm.

The reason for this selective "cold intolerance" is not glandular, as it is in someone with low thyroid function, but rather the result of cold-induced spasm of the small arteries. These vessels are not blocked as they are in arteriosclerosis, but go into spasm and close down when cold. As soon as the extremities are rewarmed, the arterial spasm breaks, and normal blood flow is restored.

This abnormal reaction to cold can be an isolated disorder, and not associated with any other complaints, in which event, it is called Raynaud's *disease*. When it is part of an underlying immune disease such as lupus, rheumatoid arthritis, or scleroderma, it is referred to as Raynaud's *phenomenon.*

The best approach to Raynaud's (regardless of whether it is the "disease" or the "phenomenon") is to avoid exposure to cold wherever possible. Here are some practical steps you can take in order to do that:

- Dress warmly and make sure your hands are properly covered when you go out in the cold. Wear mittens, in which the fingers can keep each other warm, rather than gloves, where every finger is on its own. In really cold weather, fur-lined shoes or boots are a must—you might even need battery-heated socks and boots. I have several patients in whom the condition is so troublesome that they wear socks and mittens in bed. Also, wear a hat to conserve body heat, especially if you're bald.
- If your car has been left out on a winter night, put gloves on before grasping the cold steering wheel.
- When hosting a cocktail party, let someone else serve the iced drinks. And in a supermarket, ask the clerk to remove the frozen foods you're buying from the freezer, or put your gloves on to do it.
- Buy a dishwasher (although that won't help the tip of your nose).
- Keep your fingers lubricated with any commercially available moisturizing skin cream.
- Avoid beta-blocker drugs if at all possible, because they can provoke spasm of the small arteries and further aggravate the Raynaud's symptoms. Ergotamine should not be used either; it aborts migraine headaches by constricting the dilated scalp arteries, but it has the same effect on other arteries too, and can worsen a Raynaud's condition. If you have migraines, use *Midrin* (isometheptene) instead.

If these measures do not control your Raynaud's, you will need something to dilate the affected blood vessels. The best and only drugs available for that purpose are the *calcium channel blockers* of which there are several available. I have found nifedipine (Procardia, Adalat) to be the most effective among them. You'll need a 30 mg tablet of the extended-release form (Procardia XL) every morning. Don't expect complete relief; you'll still have to come in from the cold, but this drug definitely helps. Its side effects are relatively minor and consist of facial flushing, headache, and swelling of one or both legs.

RHEUMATIC FEVER: It Licks the Joints and Bites the Heart

When I was a medical student, acute rheumatic fever was very common, especially among the young and the poor. The typical attack followed a strep throat, and consisted of: painful *arthritis* flitting from one large joint to the other—the knees, ankles, elbows, wrists; "*chorea*"—jerking movements of the limbs, lack of coordination, facial grimacing, disjointed speech; a characteristic *rash* with wavy margins; and *fever.* The diagnosis was clinched by a series of blood tests. Many parents dismissed these symptoms when they were mild in the mistaken belief that they were only "growing pains" (whatever that was supposed to mean). Years later, heart murmurs indicating rheumatic deformity of the heart valves could be heard.

We don't see as many new cases of acute rheumatic fever anymore in nations and communities where the diagnosis can be made and penicillin is in plentiful supply. But this disease remains an important problem among the poor and medically neglected everywhere.

The key to the successful prevention of rheumatic fever is recognition and treatment of the strep throat. That's a diagnosis virtually anyone can make. The throat is red, covered with white patches, and very painful. The best therapy is an immediate intramuscular shot of 1.2 million units of long-acting penicillin (children under five years of age require half that dose). In the presence of penicillin allergy, the alternative therapy is erythromycin by mouth (any preparation other than Ilosone), 1,000 mg daily in 4 divided doses for 2 weeks. Erythromycin is as effective as penicillin but not nearly as convenient, because you've got to make sure that every single dose of the antibiotic is taken—a responsibility from which the penicillin shot frees you.

If for some reason the strep throat goes unrecognized and untreated, rheumatic fever does not always follow. The problem is there's no way of predicting when it will and when it won't. However, once rheumatic fever has developed, the best we can do is control its symptoms—*there is no cure!* The first objective of treatment is to lower the fever and ease the joint pains. That requires *lots* of aspirin, at least 12 or more a day for teenagers and adults, and half that number for young children. This high dose usually needs to be continued for 2 months, because that's how long the fever and other symptoms of acute rheumatic fever otherwise last.

If there is evidence of cardiac involvement, like the appearance of a new heart murmur, then a steroid hormone like Prednisone should be added for about 4 weeks, although I'm not convinced that this actually prevents the eventual malformation of the heart valves. Such distortion often eventually requires surgical correction or replacement with prosthetic valves.

Children suffering from acute rheumatic fever feel miserable and hurt all over. Should they be kept in bed, or encouraged to stay active? Obviously, sports are out for kids with cardiac involvement, at least until the heart symptoms clear up, and that may take *many* weeks. However, if there is no evidence that the disease has affected the heart, then 2 or 3 weeks of home rest is usually enough, followed by a gradual return to normal activities.

Remember this about rheumatic fever: *One bout leaves you vulnerable to another with every re-exposure to streptococcal infection.* You should have an intramuscular shot of 1.2 million units of long-acting penicillin *every month* for 5 to 10 years after the initial attack. This is *especially* important if the heart was involved, which, fortunately, doesn't always happen. If you prefer the oral penicillin, then take 250 mg of penicillin VK twice a day, *every day,* for the same 5-to-10-year period. In cases of penicillin allergy, use erythromycin, 500 mg a day, which is half the dose necessary for the treatment of the acute sore throat.

After 10 years, you may stop the prophylactic antibiotics, but resume them if you find yourself in the midst of a strep epidemic in your community, or when you're having *any* dental work done (and that includes "simple" cleaning of the teeth) or undergoing any invasive procedure such as a D&C or colonoscopy, when bacteria can be introduced into the bloodstream—and onto your rheumatic heart valve. The result is subacute bacterial endocarditis (SBE), which can be fatal if untreated.

RHEUMATOID ARTHRITIS: Oh, My Aching Joints!

Arthritis is not a single disease. It simply means inflammation of the joint and can have various causes. Knock your knee or elbow into something hard and you end up with a *traumatic* arthritis; come down with gonorrhea or hepatitis and you may develop an *infectious* arthritis; after years of wear and tear on weight-bearing joints, many older people suffer from *osteoarthri-*

tis; when your big toe or elbow is hot, swollen, and exquisitely tender, you may have a chemical arthritis called *gout*.

Rheumatoid arthritis (RA), which does not fall into any of the above categories, is perhaps the most serious of the many forms of arthritis. It means ongoing joint pain, deformity, and progressive loss of function for years and years. Unlike most other types of arthritis, RA is not curable either by medication or surgery, nor do we know its cause(s). Some doctors think that women who take oral contraceptives are less likely to develop it; some believe it's an infection; others are convinced it is some kind of allergic response; but most of us suspect that it results from a breakdown in the immune system. No matter which of these theories eventually proves to be correct, if you are diagnosed as having rheumatoid arthritis, you will do best in the long run if you have a *caring doctor,* a good *physiotherapist,* a competent *rheumatologist,* and a *supportive family.*

Although rheumatoid arthritis is most apparent in the joints, it affects virtually every part of your body. So there are two objectives to treatment. The first is to relieve joint pain and inflammation; the second is to slow down the progress of the underlying disease itself.

The basic drug for the control of symptoms is aspirin. Unfortunately, 14 or more tablets a day may be required to relieve the pain of full-blown rheumatoid arthritis, and that's more than most people can tolerate. Gastric irritation, intestinal bleeding, especially in elderly persons, and ringing and other noises in the ears frequently result from such high doses taken for prolonged periods of time. Until fairly recently, I used to give my patients with rheumatoid arthritis enteric coated aspirin (Ecotrin). When this produced any of the complications listed above, I would switch to salicylates like Trilisate or Disalcid, which are less likely to cause gastric irritation and bleeding. These days, however, I bypass aspirin and prescribe a nonsteroidal anti-inflammatory agent (NSAID), which is more potent than aspirin and less irritating to the gut. There are many such preparations on the market most of which require prescriptions—ibuprofen (Motrin), naproxen (Anaprox, Naprosyn), indomethacin (Indocin), sulindac (Clinoril), piroxicam (Feldene), flurbiprofen (Ansaid), diflunisal (Dolobid), and diclofenac (Voltaren), to name but a few. Their dosage varies, and some are longer acting than others. But remember this—if one doesn't work, another might. The variations among them may be enough to make a therapeutic difference.

Generally speaking, all the NSAIDs cause the body to hold on to its salt. So if you have congestive heart failure or retain fluid for any other reason, an NSAID may worsen matters. Some also hurt the kidneys, others the liver, and virtually all of them can give you a skin rash and raise your blood pressure.

My two favorite NSAIDs are Voltaren, the most widely prescribed among these drugs, and Naprosyn. The usual starting dose of Voltaren is 50 mg, 3 times a day after meals; it requires a prescription. There have been reports of liver damage from Voltaren, so have your liver function checked (with a blood test) every month or so as long as you continue to take it. If Voltaren doesn't work, I switch to Naprosyn 375 mg, 3 times a day after meals, or 500 mg, twice a day.

If you require large doses of aspirin or any NSAID preparation on an ongoing basis, adding misoprostol (Cytotec) will help prevent gastrointestinal bleeding and irritation. Take 100 mg, 3 times a day (but not if you are pregnant because it can induce an abortion, one of the reasons its approval was delayed in the United States). Before Cytotec became available, we relied on antacids, or one of the acid-blocking drugs like Tagamet or Zantac for such protection. If you can't tolerate the Cytotec (some patients develop diarrhea), that's what you should take instead.

It's a great temptation to look to narcotics or steroids for relief when you're in constant pain. Don't! "Just this once, doctor" may leave you "hooked" for life. Narcotics do make you feel better, but you will pay a stiff price in the long run. Anyone who uses them on an ongoing basis faces the specter of addiction. Steroids, have their own complications, which include an increased risk of osteoporosis, peptic ulcers, high blood pressure, and personality changes.

There are several drugs that may slow down the progress of RA. They all work differently and no one is sure why or how they do. Some of these "disease modifying agents" are toxic, and their "cure" may be worse than the rheumatoid arthritis itself. But if the RA is so bad that you need more and more pain killers, and in addition feel really lousy, weak, and tired and have a low-grade fever, then try maintenance therapy with sulfasalazine (Azulfidine). Although its main use is to treat inflammatory bowel disease (ulcerative colitis, Crohn's disease), it is also effective in patients with rheumatoid arthritis. I think it's safer than any of the other "disease modifiers" described below.

Your other options, if Azulfidine fails, include hydroxychloroquin (Plaquenil), but this agent can cause visual and color perception problems, so have your eyes checked at least twice a year while on this drug. It usually takes at least 6 months for Plaquenil to work. Have patience.

Gold by injection or, more conveniently, by mouth, is another drug for you to consider. By either route, it requires very careful monitoring, so make sure that whoever prescribes it for you has had experience in its use. Serious side effects are rare, but gold can cause skin rashes, blood abnormalities, and kidney problems. The maximum dosage for oral gold is 6 mg a day.

If you've tried gold without success (or you can't tolerate it) and your arthritis is debilitating, ask your doctor about *penicillamine* (Cuprimine, Depen). Unfortunately, like gold it too has potential toxicity. It can damage the kidneys and the bone marrow, leaving you either anemic or deficient in several blood cell components.

The latest addition to the list of drugs for rheumatoid arthritis is methotrexate (Rheumatrex), which is used primarily against cancer. I have seen dramatic response in a few of my own patients who have taken it for RA. Its main drawback is potential liver damage, so make sure to have the necessary liver function blood tests while you're on this agent. But its effectiveness can be so striking and long lasting that many rheumatology specialists are beginning to use it much earlier in the course of therapy—even before Plaquenil, gold, and penicillamine.

Not infrequently, in addition to drug therapy and physiotherapy for RA, you may need injections of steroids directly into the joints, or even an operation to improve joint mobility. These can help when all other measures have failed. All in all, RA is a lousy disease to have.

SALT SENSITIVITY: Substitutes That Help Keep You Away from the Real Stuff

The taste for salt is acquired—we aren't born craving it. We learn to love it as babies when mother, who prepares our food, seasons it to her taste, something she in turn acquired from *her* mother!

Salt is a problem for many people with various ailments. For example, it is important to avoid an excess if you have *salt-dependent* high blood pressure (but that does *not* include everyone with hypertension); if you retain fluid (the most important causes of which are heart failure and various phases of the menstrual cycle); and in certain hormonal disorders. But when all is said and done, I believe that far too many people abstain from salt for no good reason.

If all your life you've eaten as much salt as your heart desired and you *must* now curtail it, your food is going to taste terribly bland. In that case, you should be aware of several options that may add some zest to your diet. I've tried most of them in order to be able to advise my own patients. The salt substitute I prefer is Adolph's (but Morton's is pretty good, too). Since almost all of these preparations contain potassium in place of the salt, they should be avoided in the following circumstances:

- If you are a diabetic taking insulin;
- If you're receiving a diuretic that contains either additional potassium or some agent that causes your body to retain it;
- If you're taking an ACE inhibitor—captopril (Capoten), lisinipril (Zestril, Prinivil), or enalapril (Vasotec)—because of high blood pressure or congestive heart failure or weakness. These medications have a potassium-sparing effect that can raise its level in the blood.

If salt is strictly verboten and you have also been advised to avoid potassium, you can learn how to enhance the taste of your food by reading *Cooking Without a Grain of Salt,* by Elmer W. Bragg. This volume contains many recipes to titillate every taste without resorting to salt. Gaylord Hauser has also prepared some very good herb and spice products that are sold in health food stores and that have neither sodium nor potassium. My favorite is Vegit; it's subtle and its taste won't dominate the dish you're cooking. If you like a barbecue flavor, try Select, a very pleasant barbecue spice. What is popcorn without butter and salt? Well, here's some good news. I'll bet that if you add some nutritional yeast to your popcorn, you won't be able to tell the difference between it and salt. I remember eating some of this popcorn in a movie theater in Colorado where yeast is often used as a salt substitute.

My friend Liz Fisher knows a lot about "heart healthy" foods, and is very imaginative to boot. When I asked her for an herbal seasoning that I could prepare at home, she came up with a recipe that I found delicious on red meat, chicken, fish, vegetables, and legumes. Combine equal parts of dehydrated onion flakes (or powder), garlic flakes (or powder), celery seed, parsley, lemon juice, and chili (hot pepper) sauce. But do your own experimenting, too. Let your imagination run wild. Don't limit your horizons to garlic and lemon. Work in some dill, horseradish, and flavored vinegars. Concoct all kinds of combinations with these and other herbs until you find one or more that you love. Don't deprive yourself of mustard either. Most commercial preparations, like Gulden's, contain only a very small amount of salt, nothing to speak of. And believe it or not, there are even very low sodium Worcestershire sauces too. Salt? Who needs salt?

In summary, you needn't go through life deprived of the pleasures of eating just because salt is not good for you. There are enough herbs and spices around to make your food taste really good.

SINUSITIS: Antibiotics Are the Best Treatment

A chronic yellow or green discharge from the nose, swelling around one or both eyes, tenderness in your face, headache that is aggravated when you press on certain points over your face, congestion of your nose, bad breath, pain in your upper teeth, low-grade fever that comes and goes, a feeling of pressure in your head when you bend over, a constant dripping at the back of your throat that makes you cough—all add up to sinusitis. An X ray of the sinuses will confirm the diagnosis.

The best way to treat sinusitis is with antibiotics. But since the sinuses are "locked in" by the facial bones, these antibiotics have a tough time getting at the infection. The most effective one, in my experience, is amoxicillin, 500 mg, 3 times a day—unless, of course, you're allergic to penicillin. If the sinus infection is chronic, as it often is, continue the amoxicillin for a full 6 weeks. If your doctor has trouble eradicating the infection because of nasal polyps or a deviated septum (the internal nose bone that separates it into a left and right side is crooked), then the polyps may have to be removed and the septum straightened.

In persons sensitive to penicillin, I prescribe Bactrim DS, 2 tablets a day for 10 days, unless they're allergic to sulfa, too. If neither the amoxicillin nor the Bactrim clears the infection, I usually go on to cefuroxime (Ceftin).

Proper antibiotic treatment can and will cure most cases of sinusitis. Don't waste your time on cough medicines, decongestants, and antihistamines; they won't do a thing for you. If your sinusitis keeps recurring, and you repeatedly require prolonged courses of antibiotics, then the sinus openings should be enlarged right there in the doctor's office in order to allow the accumulated pus to drain.

SPLINTERS: Removal Is Not Major Surgery, But . . .

When I was a kid, one of my big hang-ups was splinters. I was sure that one of them would somehow make its way deep inside my body. I was also in terror at the sight of my mother approaching with her sewing needle, recently charred over a wooden match, determined to "operate." Thank goodness, her eyesight and steadiness of hand remained unimpaired, at least until my wife could take over years later. But now, I have good news for all you children, lumberjacks, and carpenters who are as chicken about splinters as I was and still am. There is an effective way to remove them—one that doesn't hurt and that you can do yourself without a needle. I read about this method in a journal of dermatology last year, and I salute Dr. R. Copelan for this major contribution to cowards everywhere! All you have to do is buy a Mediplast patch, which contains 40 percent salicylic acid, and cover the splinter with it. The acid softens the most superficial layer of the skin under which the splinter is usually burrowed so that it literally just lifts off in 2 or 3 days. Those that are more deeply embedded take a little longer. These same plasters also dissolve warts, calluses, and corns. Salicylic acid in its weaker strength is an antidandruff and antidermatitis agent.

I hesitate to advise my patients to use a treatment unless I have personally tried it on myself wherever possible. I was anxious to evaluate Mediplast after learning about it. However, I just wasn't able to get a splinter for several months, no matter how hard I tried. But then last winter, while picking up a log for my fireplace, I got "lucky." A splinter lodged in my finger! It's the first time in my life such an injury made me happy. I rushed down to the drugstore, bought a couple of Mediplast patches, put one on the injured finger—and waited. *Voilà!* Just as Dr. Copelan said it would, the splinter came out in 2 days—no needles, no pain, no threat of infection, and no fuss. So, if you ever get a splinter again, the best way to treat it is with topical salicylic acid—unless, of course, you enjoy the old way!

STROKE: Not Always Hopeless

Don't confuse a stroke with a heart attack. A stroke is what happens when the blood supply to a portion of the *brain* is interrupted; a heart attack is what you suffer when the blockage is in an artery within the *heart*. Since the brain controls speech, movement, sensation, intelligence, and a host of other vital bodily functions, whether or not one survives a stroke, and in what condition, depends on the mechanism by which the blood supply to the brain was cut off, the size of the artery involved, and the specific area of the brain that was damaged.

The flow of blood in a cerebral (brain) artery may be interrupted either abruptly or insidiously, and it can occur in several different ways. The most common mechanism is arteriosclerosis, which leads to gradual narrowing of a blood vessel. When the vessel finally closes, a stroke ensues. A stroke may also be the consequence of hemorrhage from the sudden bursting of a brain artery weakened over the years by the incessant pounding of high blood pressure. Finally, a clot originating elsewhere in the body can make its way to the brain and end up blocking one of the arteries deep within it. Such emboli can arise in the heart, where they form after a heart attack or in certain kinds of heart valve disease like mitral stenosis, when the heart rhythm is out of synch (atrial fibrillation); or the emboli may break off from one of the vessels in the neck (the carotid arteries). Closure, hemorrhage, and embolus account for the vast majority of strokes, but they can also be caused by rare blood diseases, brain tumors, and spasm.

The best way to deal with a stroke is to prevent it. So, if you're prone to arteriosclerosis because of a bad family history, high cholesterol, elevated blood pressure, or heavy cigarette smoking, you should work hard at correcting as many of these risk factors as you can. To avoid a brain hemorrhage, keep your blood pressure normal, and if a potential source of a traveling clot is found, blood thinners can save your life.

You may have a warning before a stroke occurs—a phenomenon called a *transient ischemic attack (TIA)*. You see double for a few minutes, or lose vision in one eye, or experience temporary weakness or paralysis of a limb, or loss of speech—all of which clear as suddenly and as mysteriously as they came on. If they do, don't simply congratulate yourself on your good luck and go about your business as usual, hoping that you'll never have a recurrence. Such TIA's should be treated immediately with 1 or 2 aspirins a day for the rest of your life to prevent a full-blown stroke. Some doctors prefer anticoagulation with blood thinners like Coumadin, but in my opinion, aspirin is usually enough.

You may not be lucky enough to have the warning of a TIA, and instead suddenly sustain the real thing—which can manifest itself in paralysis, blindness, speech difficulties, which do not clear up. The first thing to do under these circumstances is to get to the hospital immediately. No one can predict how a stroke will end. For example, slurring of speech may progress in hours to an inability to swallow or breathe. If you're in a hospital during that critical time, you can often be helped over such a crisis. You've got no chance at home.

The majority of stroke victims do improve, and sometimes all damage ultimately clears completely. That happens when a stroke is due to a hemorrhage and the blood is re-absorbed, or when an obstructing clot shrinks and the residual brain damage is not as great as originally feared.

While the doctors are watching to see how far and how fast the stroke is progressing, you may immediately be given intravenous heparin (a rapidly acting anticoagulant that is used *if* the stroke is not due to hemorrhage but to a blood clot or an embolus). That diagnosis can now be established by a CT scan of the brain, which almost every stroke patient must have soon after admission to the hospital. After several days of heparin therapy, you may require ongoing oral anticoagulation with Coumadin if the cause was an embolus or if you have chronic atrial fibrillation. In addition to blood thinners, you may also temporarily need mannitol to reduce the swelling that often occurs in the brain during the early stages of a stroke.

If you are left paralyzed or unable to speak, physiotherapy can often restore some function even when the outlook seems hopeless. But the success of that treatment requires expert therapists and your cooperation. I'm always amazed at how much a crack physiatry team can do for a severely impaired patient using a combination of perseverance and modern technology.

There is active ongoing research in the treatment of stroke, and new frontiers are being established. For example, some researchers are now inserting balloons into arteries that have closed. This angioplasty technique is similar to that done in the heart and the legs. New medications are also being developed to dissolve clots within the brain just as they are for the heart and legs.

In summary, at this time the bottom line is to prevent strokes by identifying and correcting your risk factors. Should you develop a TIA, you must take aspirin (unless there is some reason not to do so). If a stroke comes on, get to a hospital right away so that you can be protected during its evolution and receive treatment for any life-threatening complications. Anticoagulation (blood thinning) is the hallmark of treatment except if your stroke was due to a hemorrhage into the brain. Most persons who suffer stroke recover completely or very nearly so. Those who do not can often be improved by sophisticated physiotherapeutic approaches.

STYE: A Matter of the Right Antibiotic

A *stye*—the red, swollen, tender nodule under your eye—is annoying and painful, but not life-threatening. It develops when staphylococcus bacteria that normally inhabit the skin infect a wax (sebaceous) gland in the eyelid. The best treatment is warm packs—with or without salt or boric acid—for 30 to 60 minutes, three times daily, if possible. In addition, I recommend Bacitracin, an antibiotic ophthalmic ointment, twice a day. This regimen will usually cure the stye in about 2 weeks. Neosporin (ophthalmic), an ointment that contains several topical antibiotics, is as effective as Bacitracin, but is not my first choice because it is more likely to result in a local allergic reaction.

SUNBURN: You Should Have Stayed in the Shade!

The impressive statistics to which I referred in the section on burns do not, of course, include the countless sunburns, usually self-inflicted, that occur every year. In Victorian England, women went to great pains to avoid the sun because "pale" was considered to be very feminine and a sign of good breeding. A tan, in both sexes, was in those days the hallmark of the laborer exposed by necessity to the sun. By contrast, until quite recently in this century, a bronzed skin conjured up images of affluence—lots of leisure time, exercise, sun, and fresh air. It was viewed as both beautiful and healthy, in men and in women. We now know that that's all wrong. The long-term consequences of chronic overexposure to the sun are: (a) skin cancer, of which one type, the malignant melanoma, is life-threatening, and (b) skin damage with resultant premature aging and wrinkling. But despite all the public education about the hazards involved, countless diehards still spend most of their vacation lying motionless for hours baking in the sun by the pool or on the beach. When they can't get the real thing down South, they frequent tanning parlors up North.

If you love the feel of the hot sun on your skin and choose to be exposed, then at least heed the following precautions. Don't sunbathe between 10:00 a.m. and 3:00 p.m., when the rays are strongest. Apply a sunscreen with a sun protective factor of at least 15; the lighter your skin, the higher that number should be. (If you are dark-skinned and tan easily, you may get away with a number 8.) The higher blockers allow you to remain in the sun for a longer time. Remember that one application doesn't last all day. Your skin must be "touched up" every 30 to 60 minutes, especially if you've been in and out of the surf or pool, or perspire excessively.

If your sunburn hurts, take some aspirin and apply cold compresses or sit in a cool bath. If you're still uncomfortable, use a topical steroid like Cortaid (available without a prescription) where it pains most. If you blister, you can drain the fluid with a sterile needle, but leave the collapsed blister tissue alone—that is, don't peel it away. Polysporin ointment a couple of times a day on exposed bare skin will help prevent local infection. Unless you've been shipwrecked, marooned, or have been floating on a raft without any protection for several days, the simple measures described above are all you'll normally need to take care of a sunburn.

If you used to be a sun worshipper, even though you *now* know better, see a dermatologist at yearly intervals to check for melanomas and other skin cancers. They can appear many years after exposure. And if you happen to see any wart or pigmented "thing" on your skin, especially if it's brownish-black with an irregular jagged border rather than a smooth one, have your skin examined—fast.

Premalignant eruptions are often present on the skin in people who have been exposed to too much sun over the years. Most dermatologists use 5-Fluorouracil twice a day for several weeks to remove these potentially cancerous spots. This treatment is effective, but can leave you hurting for a couple of weeks.

In addition to skin damage from too much sun, there is also the matter of sun *allergy* even after only minimal exposure. Women using cosmetics with specific fragrances are particularly vulnerable to developing blisters and dark spots after being out in the sun for only brief periods. So is anyone taking tetracycline, or a derivative such as doxycycline (Vibramycin), an antibiotic used by many tourists for the prevention and/or treatment of traveler's diarrhea. Several diuretics, as well as Benadryl, a widely used antihistamine, can also lead to skin damage in persons exposed to sunlight.

If you've enjoyed sun bathing over the years and your facial skin is now dry and wrinkled, all is not lost. Unless the wrinkles are too coarse, they can be improved by Retin-A, in a 0.1 percent cream (doctor's prescription is required). You must keep *totally* out of the sun for the duration of this treatment, and don't be surprised if the cream leaves your skin reddened and uncomfortable for a few days. That's to be expected. I have seen some really impressive improvement in the appearance of several of my patients who have tried topical Retin-A.

In summary, here's the bottom line with regard to sun bathing. The "beauty" it confers is only skin deep, but the problems it can cause are far from superficial. You will pay dearly for whatever cosmetic benefit you derive early in life (and then some) should you end up with skin cancer or prematurely aged, washboard skin at a time in your life when most of us prefer to look younger, rather than withered.

SYPHILIS: Serious If Not Cured

Although one doesn't hear much about syphilis anymore, what with all of the concern over AIDS, this disease is still with us. Someone with gonorrhea, chlamydia, or any other sexually transmitted disease should always routinely be examined and treated for syphilis because all these infections are handmaidens to one another. And treatment should be started immediately in order to avoid the late complications of this disease.

Actually, believe it or not, you can breathe a sigh of relief if after some sexual indiscretion you find that you have contracted nothing more than syphilis. It is neither permanent like herpes, nor ultimately fatal like AIDS.

The first evidence of syphilis is the painless "cold sore" that appears about 3 weeks after exposure, and vanishes without a trace after several days or weeks. Such "lesions" develop at the site of sexual contact—usually the genitalia, anus, or mouth, depending on who does what to whom. Once the cold sore clears, the infecting organism moves into the interior of the body and settles down in virtually any organ.

If the acute attack was undiagnosed and untreated, the disease enters a *secondary* stage usually 6 to 8 weeks later. Patients feel as though they have the flu; their glands enlarge throughout the body, which is covered by a pink rash.

The best treatment for syphilis within one year of infection is Benzathine penicillin G, 2.4 million units by injection. If you are found to be infected with syphilis in the course of a routine blood test, and were never treated and aren't sure when you acquired the disease, you're assumed to have the *tertiary* form. You'll now need 3 of these penicillin shots at intervals of 1 per week. If you are allergic to penicillin, there are several excellent alternatives. I prefer doxycycline (Vibramycin), 100 mg by mouth, twice a day for 15 days.

Everyone with syphilis (especially homosexuals) should also be tested for AIDS a few months later, because the same contact may have given you both diseases.

TETANUS: The Best Way to Keep Your Jaw from Locking!

Tetanus is nicknamed "lockjaw" because it can result in such severe muscle spasm that the jaw sometimes cannot be opened and closed. This infection is caused by an organism that thrives in the earth and loves to wallow in animal feces. So you're at risk for tetanus when you get a dirty wound or break in the skin, especially if it happened in the country near a barn. (A surgeon who has just finished scrubbing and cuts himself with a sterile scalpel needn't worry about contracting tetanus.) Once the bug enters your body, it secretes a nerve toxin against which there is no effective antidote, and which produces widespread muscle spasm, particularly in the muscles that control respiration. So patients don't die from locking of the jaw, but rather from the inability to breathe.

The key to dealing with tetanus is to render yourself immune by the vaccination so that if and when the organism does gain access, it will be overwhelmed by your antibodies. Exposed persons without such immunity should receive an immediate injection of antibiotics before the tetanus bug has been able to release any substantial amount of its toxin.

Given the possible fatal consequences of tetanus, what's the best way to protect yourself against what appears to be an "innocent" laceration? Here are my recommendations. No matter how or where your skin was broken, if you were fully immunized in childhood, and have been getting regular tetanus boosters every 10 years as you're supposed to, simply clean the punctured area with an antiseptic (I like Betadine cream). If the injury occurred in suspicious circumstances such as stepping on a rusty nail or falling in a stable, then also get a shot of 1.2 million units of penicillin as soon as possible; that will kill the tetanus bacteria. If you were vaccinated in childhood but *have not had a booster* in the past 5 or more years, you should have an injection of tetanus *toxoid* together with the penicillin. If you're not sure whether you *ever* received any immunization (many new Americans come from countries where vaccination was not routine), then you must assume that you're vulnerable and should have *immediate* protection. That means penicillin, *antitoxin,* and an injection of *human immune globulin* as close as possible to the wound.

Because tetanus vaccination and follow-up booster doses at regular intervals are so important, let me review the *immunization schedule* for you. Every infant should be vaccinated at about 2 months of age, followed by a second shot 2 months later, and a third one 2 months after that. The fourth dose is given 15 months after the third dose, and a booster just before entering school at 5 or 6 years of age. From then on, it should be administered every 10 years.

THYROID HORMONE IMBALANCE:
Resetting Your Body's Thermostat

The thyroid gland in your neck regulates your internal thermostat with exquisite sensitivity by secreting exactly the right amount of thyroid hormone. Should it produce too much or too little, all kinds of bad things begin to happen. A deficiency of thyroid—*hypothyroidism*—leaves you feeling, looking, and acting like a "zombie": Your hair falls out, you gain weight, you feel cold, and you're constipated. On the other hand, an excess—*hyperthyroidism*—makes you jumpy, irritable, and overactive: You lose weight, your eyes may bulge, you may develop an irregular heartbeat (like President Bush), and you always feel overheated. Either situation can be corrected.

Hyperthyroidism

The amount of the hormone manufactured by the thyroid gland and the rate at which it is discharged into the bloodstream depend on a complicated feedback system controlled by the pituitary lying deep within the brain. This gland produces a protein called thyroid stimulating hormone (TSH)—which in turn stimulates the thyroid gland to make its thyroid hormone in normal amounts. Sometimes, and more frequently in females than in males, the immune system produces an antibody that acts like TSH. The thyroid gland is then whipped into a frenzy and produces an excess of its hormone. This overabundance has an impact on virtually every biological function. The end clinical picture is one of a nervous individual with a rapid pulse, high blood pressure, profuse sweats, heat intolerance, diarrhea, weight loss (despite a ravenous appetite), fine silky hair, popping or bulging eyes, on rare occasions an increased blood sugar, and often some enlargement of the hard-working thyroid gland in the neck.

There are three different ways to stop overproduction of thyroid hormone. You'd think that the most logical approach would be to prevent the immune system from making the stimulating antibody that is the villain of the piece. You're absolutely right, except that there is no safe way to do that. So we are left with the following options:

- We can remove the thyroid gland surgically so that the antibody has nothing to stimulate;
- Instead of removing it we can burn it away with a drink of radioactive iodine, which is what Mr. Bush was given;
- We can administer medicine that enables the thyroid gland to ignore the stimulating signal and produce only as much hormone as is needed.

My preference among these three options for *almost* all of my hyperthyroid patients is radioactive iodine. The only (and very frequent) complication you can expect is to be left permanently hypothyroid. No matter how expert the doctor who determines the dosage of the radioactive iodine, almost every patient receiving it ultimately ends up with an underfunctioning gland. So what? You simply take 1 or 2 thyroid pills daily to replace what is missing, and you can lead a perfectly normal life.

But radioactive iodine is not for everyone. Here are the exceptions: If you're under twenty years of age, or are pregnant, there is a *theoretical* possibility that the radioactivity in the drink may cause cancer years later or induce genetic abnormalities in your fetus. Mind you, I have never encountered either of these complications in my own practice, but one can't be absolutely sure. Virtually everyone else, however, should in my opinion choose this therapy, especially those who are allergic to the drugs that suppress the overactive thyroid gland or those who have underlying heart disease, and in whom the risk of surgery is increased.

My second choice for the treatment of hyperthyroidism is drug therapy to block production of the thyroid hormone by the gland. The best agent for this purpose is propylthiouracil (PTU). This approach is not nearly as convenient as radioactive iodine, where you drink a small amount of a tasteless liquid once, and that's all there is to it. PTU must be taken every 6 to 8 hours for months on end. Like any drug, it has potential side effects—rash, fever, joint pains, and blood abnormalities. Its biggest drawback, however, is that less than half of the patients so treated are permanently cured. However, if you are pregnant, this type of antithyroid therapy is the route to go.

Surgery is my last choice, but it is sometimes required. For example, if you have a *very* large thyroid gland that has begun to resemble a bicycle tire around your neck, causing mechanical pressure on nearby structures, neither the antithyroid drugs nor the radioactive iodine will eliminate it. They may reduce the size of the thyroid somewhat, but surgery is the only way to get rid of it. But such surgery is not as simple as an appendectomy or even a gallbladder operation. It should only be performed by a surgeon who has had lots of experience with this particular procedure. It is tricky, delicate, and if done less than perfectly, can be ineffective. For example, enough thyroid tissue may be left behind to cause your symptoms to persist. Even more important, vital nerves in the area of the thyroid gland might be injured accidentally, leaving you hoarse and with a paralyzed diaphragm in your chest. The parathyroid glands, which lie very close to the thyroid gland and which control calcium metabolism, can be inadvertently damaged, too. Although surgery is usually curative, 20 percent of patients do have a recurrence of their hyperthyroidism within 2 years. Also, as with radioactive iodine, the operation is apt to leave you hypothyroid and you will need to take thyroid hormone supplements forever.

Hypothyroidism

Hypothyroidism is not diagnosed often enough. There are so many people around who are tired, listless, chronically constipated, overweight, whose hair falls out, whose menstrual flow is too heavy, whose skin is dry, and who think of every possible explanation for their symptoms except the right one. They spend tons of money at weight-reduction clinics, on diet books, creams to moisten their dry skin, laxatives to get them going in the morning, vitamin shots to restore their energy, and iron supplements for their "tired blood."

Identifying hypothyroidism requires a high index of suspicion by you and your doctor, followed by blood tests that measure how much thyroid stimulating hormone (TSH) is being produced by the brain and how much thyroid hormone is actually circulating in your body. As long as your blood is being drawn anyway, also have it checked for cholesterol, which is likely to be high when thyroid function is low.

Once the diagnosis of hypothyroidism is made, you'll be prescribed supplemental hormone. Unlike the situation with an elevated blood sugar or increased blood pressure, a dietary or weight regimen will not make any difference—you *must* replace the missing hormone. But that doesn't mean that you go to the drugstore and buy whichever preparation your druggist offers you—or for that matter, the cheapest product available. Not at all. First you need a doctor's prescription. You cannot obtain thyroid supplements over the counter in the United States. Then, there are three replacement choices. The first is the "natural" product for which most people instinctively opt because of the word *natural*. This particular preparation, made from beef and pork thyroids, is effective, and less expensive than the synthetic forms. However, I don't prescribe it because I think potency varies somewhat from batch to batch, especially among the generic brands. That leaves two synthetic thyroid hormones, levothyroxine (Synthroid) and liothyronine (Cytomel). I prefer Synthroid because it has a longer duration of action, which results in more stable replacement levels. But since Synthroid takes several days to work, I start therapy with Cytomel, which has a more rapid onset of action, together with the Synthroid. After a week or two, I stop the Cytomel and leave the patient on Synthroid alone. The usual replacement dose is 0.1 to 0.2 mg, but I start with less (0.05) and increase it every 3 weeks until the target dose is attained.

Once begun, thyroid hormone is usually continued indefinitely. It is not like an antibiotic, which you stop when the infection is over. Hypothyroidism is never "cured." But thyroid supplements in the proper amount have absolutely no known side effects; they are safe and relatively inexpensive. However, the dose must be carefully adjusted, especially if you have any heart trouble (such as angina or cardiac rhythm disturbance). Although it's okay to take thyroid while you are pregnant because the hormone does not cross into the fetal circulation, discuss dosage with your doctor if you are breast feeding. Thyroid hormone makes its way into mother's milk and is not especially good for the infant who doesn't need it.

Because thyroid speeds up the metabolism, the hormone is popular among some misinformed and misguided persons who take it for weight loss or for "energy." Don't you do it! Thyroid hormone should only replace what's missing and not increase normal levels.

You will know you are overdosing on thyroid if you become nervous and irritable, experience palpitations, develop an irregular heart rhythm, have more angina, can't sleep, or get the shakes. But if you are truly hypothyroid, the right amount will make you feel like a new person: Your skin will soften, your energy level will increase, your weight will go down, cold weather won't bother you nearly as much, your bowels will begin to move regularly, your periods will become predictable once more, your cholesterol level will drop, and if you have been infertile, you're apt to become the proud parent of a bouncing baby.

TOURETTE'S SYNDROME: Damn You, Damn You, Damn You

There are more than one million Americans, and some of their relatives, who share an uncontrollable compulsion suddenly to jerk about or say things they shouldn't. You may have spotted those with a mild form of this disorder by their tics, frequent shrugging of the shoulders, the grunting noises they make, even their explosive barks. Aside from occasional embarrassment, most of these individuals function fairly normally. But the really seriously afflicted—those who cannot control their loud and usually obscene utterances—are, for the most part, socially ostracized and kept out of sight and under wraps. This compulsive behavior is called *Tourette's syndrome,* named after the French doctor who first described it in the 1800s.

Tourette's is a genetic abnormality that runs in families and often includes other forms of abnormal behavior. It's three times more common among males than females, and although it is present in all races, the incidence is higher among whites.

You may find some of the symptoms amusing as an onlooker, but they are a real tragedy to someone who has them. Remember the story of the traveling salesman who had an uncontrollable wink? He arrived in Boston one day, got into a cab at the airport, gave the driver a big tip, and told him he had no hotel reservations. "Take me where I can get a good night's sleep," he said with a wink. The driver pocketed the tip, winked back—and took the salesman to a house of ill repute.

When I was appearing with Gary Collins on the television program "Hour Magazine," we decided, alas, to invite a young man with Tourette's syndrome on the program—to demonstrate a problem of which most people are unaware. We hoped to be able to get through the interview without our guest's exploding into obscenity. When he finally arrived at the studio, we were prepared for the worst. After he and I were introduced, our guest inferred my religious affiliation from my name, and after chatting for a while, suddenly, and for no apparent reason, broke into, "Kike, kike, kike, kike, kike." He was then taken to the makeup room where, while getting his face prepared for the cameras, he smiled at the young cosmetician and exploded with an enthusiastic, "Fuck, fuck, fuck, fuck, fuck." Despite our trepidation at what this man would say on camera, Gary and I felt we could deal with any expletives by "bleeping" them out. (The program was taped before an audience). Well, we were lucky because his virtually continuous vulgarities were uttered under his breath and barely audible. Those that weren't were obliterated, thanks to modern technology. Despite his high motivation and equally high IQ, this unfortunate chap was a social misfit, totally incapacitated by his illness, unable to live a normal life or even to attend school.

In the great majority of cases, Tourette's begins early in life with repeated and involuntary tics, grunts, barks, groans, throat clearings, bodily movements, none of which have any purpose. In about 10 percent of cases, however, there are absolutely uncontrollable, unpredictable "subconscious" obscene outbursts (coprolalia) such as those of our television guest.

Tourette's is now, to a large extent, treatable though not curable with medication. Psychiatry is of no use whatsoever. There is no point in telling these people to "cut it out," threatening them, or punishing them. In fact, their abnormal behavior is usually aggravated by such additional stress. Fortunately, many cases clear up on their own after a few years, and the condition often becomes less severe with time. There are several drugs that will control the symptoms. The best is haloperidol (Haldol). It comes in ¼ mg, ½ mg, and 1 mg tablets, and the average amount required is about 2 to 3 mg a day. When Haldol's side effects are too troublesome (extreme fatigue, symptoms mimicking Parkinson's disease, personality change in children), clonidine (Catapres) is used instead. It comes in 0.1 and 0.2 mg strengths, and a total of 0.3 mg a day is usually required.

If your child has a troublesome tic, consult a neurologist. He or she may be able to reassure you that these symptoms are only a habit and will eventually clear. Even if the disorder has lasted a year or more, there's still hope that it will either improve or even disappear. Remember that Tourette's is not a psychological disorder. The condition is beyond the child's control. Neither punishment nor bribery will help.

TUBERCULOSIS: Forgotten But Not Gone

Most people under forty years of age don't give tuberculosis a second thought. They weren't around in the days when this was an untreatable and major killer. But those of us who are older remember when the only therapy for TB was good nutrition, lots of rest and fresh air, and confinement in a sanitorium in the mountains. If your case was really bad, then the affected lung was "collapsed" in order to "rest" it.

The first drug specifically for the treatment of tuberculosis became available in the late 1950s. Since then, what was once a scourge is now treated with antibiotics just like any other infection (although for a longer period of time). However, despite the availability of a specific cure, the number of cases of tuberculosis has increased dramatically in the last few years and is still on the rise! In 1989, 23,500 new cases were reported, up 5 percent since 1988. The reasons for this resurgence lie in the increased numbers of elderly whose nutrition and general health is poor, whose immune systems and ability to resist infection are impaired by virtue of their age or other disease, and who are living in crowded nursing homes. Also, there continues to be substantial immigration into the United States from countries where tuberculosis is still prevalent due to poor diagnosis, inadequate public health measures, and the lack of treatment. But another extremely important factor is the spread of AIDS, in which the body's ability to repel or deal with tuberculosis is drastically compromised. Nevertheless, tuberculosis can be contained because it is curable. We have but to remember how it is spread (the tubercle bacillus is coughed into the air by someone with the disease and inhaled by an innocent bystander close by), take proper precautions, always keep the possibility of TB in mind whenever a cough is chronic, and begin treatment as soon as the diagnosis is made.

There are three key drugs that can kill the bug. They're all administered by the oral route, no injections are necessary, but they *must* be taken faithfully as prescribed, *every single day.* You dare not be casual in your attitude. Omitting the medication now and then can lead to relapse, prolongation of the illness, and a longer treatment time.

One of the prime drugs against tuberculosis is *isoniazid* (INH), which destroys the tubercle bacillus in dosages of 300 mg a day, both for children and adults (this is one of those unusual situations in which dosage does not vary with age and weight). During the 6 to 9 months during which you must continue this therapy, be sure to be checked at frequent intervals for any evidence of an allergic reaction to the medication, and for liver trouble. Report any shooting pains in your arms or legs (the latter symptoms are due to inflammation of the nerves, which is sometimes caused by this drug). In order to reduce that likelihood, take pyridoxine (vitamin B_6) along with it.

Although isoniazid is effective against TB, it cannot do the job alone. You'll also need a specific antibiotic called rifampin (Rifadin, Rimactane) in a dosage of 600 mg a day. The most obvious side effect of this drug, which is apt to startle you unless you're prepared for it, is the bright orange color it imparts to your urine and to your tears! (If you don't want anyone to know you're being treated for tuberculosis, don't cry in public!) Like isoniazid, rifampin can occasionally hurt the liver, so alcohol should be avoided while taking both these medications. Although isoniazid and rifampin must be continued for at least 9 months, you become much less infectious to others as early as 3 weeks after you've begun taking them.

Here is more good news. If you add one other medication called pyrazinamide (it comes in 500 mg tablets, and you take 4 a day), you can reduce the length of treatment with INH and rifampin from 9 months to 6 months. Like any other drug, pyrazinamide can cause side effects—nausea, liver trouble, pain in the joints, fever, and flushing—but in my experience, most patients tolerate it well.

After you've been on the three-drug treatment (isoniazid, rifampin, and pyrazinamide) for 6 months, consider yourself cured. But sometime during the course of therapy your doctor will want to check your sputum just to make sure that it is, in fact, free of the tubercle bug.

The treatment of tuberculosis no longer requires hospitalization unless you're very sick; nor is isolation required. People in contact with patients, even those with active disease, don't wear masks anymore (which were useless anyway), or caps, gloves, and gowns; dishes need not be boiled and kept separately as they once were; laundry of patients with TB can be done in the same machine, and at the same time, as that of other family members. *The key to prevention is keeping those droplets out of the air.* So if you have

TB, make it a habit, at any stage of therapy, and even after you're cured, always to cover your mouth when you cough, and dispose of your tissues carefully. Unless you do, the tubercle bacilli explode into the air where they hang around long enough to infect anybody who is in close and frequent contact with you.

Every patient with AIDS who has a cough should be tested for tuberculosis even though most such coughs are due to *Pneumocystis carinii,* because the reduced immunity resulting from AIDS leaves one vulnerable to tuberculosis. However, even if you have AIDS, your TB can be cured with the combination of drugs mentioned above.

ULCER: Possibly an Infection!

The term *peptic ulcer* refers to a break in the lining of either the *stomach* or the first part of the small bowel that leads from it, the *duodenum.* We've come to understand a great deal more about ulcers in recent years—what causes them, how to prevent them, and how best to treat them. As far as cause is concerned, the bottom line is still stomach acid. Either you have too much acid or you have an abnormality in the cell lining of the gut that renders it unable to deal with even normal amounts. In either case, diet is rarely the culprit. But you can develop severe gastric irritation and erosion of the stomach lining (not quite the traditional ulcer) from too much aspirin or nonsteroidal anti-inflammatory agents (NSAIDs), severe burns, sudden serious stress, long-term cortisone therapy, and from excessive amounts of caffeine, alcohol, and tobacco. All of these substances interfere with the ability of the stomach lining to resist the action of whatever acid is present in its immediate environment.

Diet is no longer the mainstay of ulcer management. Instead, there are several drugs that relieve, prevent, and/or cure ulcers by the following mechanisms.

• *Antacids,* which are popped, chewed, and swallowed all day long by some people, relieve symptoms by *neutralizing or reducing the amount of acid* already present. But antacids are inconvenient because unless you take them conscientiously 4 or 5 times a day between meals, your ulcer pain will not be controlled. Moreover, many antacids have troublesome side effects. Some cause constipation, others will give you diarrhea, several contain more salt than is good for you if you also have high blood pressure or congestive heart failure, and too much antacid can give you more calcium than you should have. So antacids have largely been supplemented *but not replaced* by other drugs.

• There are several drugs on the market that *prevent the stomach from making acid* in the first place. These agents, called *H2 blockers*, all require prescriptions. Although I usually prescribe ranitidine (Zantac), they are all similar—the others include cimetidine (Tagamet), famotidine (Pepcid), and nizatidine (Axid). When the ulcer is acute or "hot," I have my patients take 150 mg of Zantac twice a day (upon arising and at bedtime). That will heal about 80 percent of all ulcers within 4 weeks. The remaining 20 percent require treatment for a full 8 weeks. After the ulcer has healed, it very often recurs. To reduce that likelihood, I advise continuing 150 mg of Zantac twice a day for 8 weeks, and then 1 tablet at bedtime for another 8 weeks.

• Another medication, *sucralfate (Carafate), physically protects the stomach from acid injury.* Once swallowed it makes a beeline for the ulcerated area, coats it for 4 to 6 hours, and then peels off—sort of like a Band-Aid. It is not absorbed by the body. Unlike the H2 blockers, Carafate is not a preventive. You must actually have an ulcer on which it can sit, and you need 4 a day on an empty stomach in order to keep the raw area covered. Carafate alone is just as effective against acute ulcers as are any of the acid-blocking drugs. When an ulcer is slow to heal, I administer Zantac and Carafate together.

• *Misoprostol (Cytotec)*, which *reduces stomach acid content and increases the resistance of the lining cells to the acid,* prevents gastric ulcers but does not cure them. This medication makes it possible for some patients who need large amounts of aspirin or NSAIDs for, say, arthritis, to continue using them with less risk of ulcer formation. I have found a dosage of 100 mg tablets, 4 times a day (1 with each meal and at bedtime) for as long as you take the NSAIDs, to be well tolerated, much more so than the officially recommended 200 mg strength, which is apt to result in diarrhea. But don't take Cytotec if you are pregnant or about to be; it can induce abortion.

• Certain *antibiotics* can *eradicate the ulcer "infection!"* Many ulcers recur within 6 months after they've been healed by conventional treatment and in at least some cases, it's because of the presence of bacteria in the ulcer itself. Although the responsible organism, *Helicobacter pylori,* may be a normal inhabitant of the gut, it can, under certain circumstances, produce ulcers, slow down the healing process, or induce recurrence. In order to prove this theory, the doctors who first made the observation (all healthy

Australian men) actually swallowed a dose of these organisms themselves (now there's dedication for you!). Sure enough, even though none of them had ever previously had any stomach problems, they all developed peptic ulcers! The best way to eradicate *Helicobacter* after its presence has been proven in *your* duodenum is to take the antibiotic amoxicillin (or erythromycin), 500 mg, 4 times a day, together with metronidazole (Flagyl), 6 of the 250 mg a day for 2 weeks, in addition to the standard ulcer treatment. Some doctors also give their patients Pepto-Bismol. Remember, drink no alcohol with the Flagyl (or the ulcer, for that matter).

A popular myth that has gone by the boards is that drinking lots of milk helps heal an ulcer. Don't do it! Not only will it put weight on you and perhaps elevate your cholesterol level, but the protein in the milk may actually delay healing by stimulating the stomach to make more acid.

If you have an ulcer, it's important to *avoid aspirin, caffeine, chocolate, tobacco, and any form of alcohol*—at least during the acute stage. No combination of drugs is going to do you much good if you continue to smoke, drink coffee, pop aspirins, or drink booze.

In summary, we have made great strides in our understanding of ulcers—there are more potent antacids, new medications that block formation of the acid, products that cover the ulcer like a Band-Aid, and one that allows you to continue taking aspirin even if you are ulcer-prone. But perhaps the most important and dramatic piece of new information is the observation that antibiotics can sometimes actually prevent ulcer recurrence.

URINARY TRACT INFECTIONS: When You Have to Go Too Often and It Hurts

If your bladder needs emptying every few minutes, and it burns like the devil when you void, or if you have a low-grade fever, feel like hell, always have the feeling that you need to "go" and then when you do, only a few measly drops come out—you have a urinary tract infection.

Such infections are ten times more common in women than in men. The reason is the difference in anatomy between the two sexes. The tube through which urine leaves the body (urethra) from the bladder is shorter in women than in men. So an infection or irritation of the urethra—resulting from vigorous intercourse or poor personal hygiene, or wiping from the anal area toward the vagina after a bowel movement, thus introducing bowel organisms into the urinary tract—is more easily spread to the bladder, where it hangs on unless treated.

Bladder infection, or cystitis—often called honeymoon cystitis—is the most frequent physical complaint of female patients. The best way to treat such cystitis is to prevent it. Every woman should drink at least 2 glasses of water before intercourse, and void immediately after it is over. Doing so irrigates the urinary tract and renders infection less likely. A dose of trimethoprim-sulfamethoxazole (Bactrim DS, Septra DS) shortly after sex may also reduce the frequency of urinary tract infections, and is worth a try unless, of course, you're allergic to sulfa.

The symptoms of cystitis are usually due to *E. coli,* an organism sensitive to sulfa. Once symptoms have appeared, I advise my patients to continue to drink lots of water, avoid alcohol, and take 1 sulfamethoxazole (Bactrim DS) tablet twice a day for 10 to 14 days. For those who are allergic to sulfa, I prescribe cephalexin (Keflex), 250 mg, 4 times a day for 10 days, except in cases of penicillin allergy. (Although Keflex is not actually penicillin, there is some chemical overlap between the two.) A culture of the urine should be obtained when symptoms first appear, just to confirm the fact that it is sulfa-sensitive, but repeat analyses are rarely necessary.

As a woman approaches menopause, the vaginal walls become dry because of a deficiency of female hormone (estrogen). This loss of moisture leaves the area prone to urinary tract infection, especially after the friction and trauma to the vagina associated with sex. Estrogen replacement and vaginal estrogen creams help prevent this complication.

Urinary Tract Infections in Men

Repeated urinary tract infections in men are most often the result of an enlarged and infected prostate gland (prostatitis) that is interfering with the complete elimination of urine from the bladder and also causing cystitis. The residual urine just sitting there is a great place for bacteria to socialize and multiply. The symptoms in men are similar to those experienced by women—they need to "pee" very frequently and urgently several times a night, and have the sense that they have never quite emptied the bladder.

Before starting treatment for a urinary tract infection, your doctor may elect to obtain a urine *culture* to identify the organism causing the trouble.

Sensitivity tests are then applied to this specimen to discover which antibiotic will be most effective against it. As is the case in women, a bacterium called *E. coli* is the usual culprit in such urinary tract infections, and it is almost always vulnerable either to Bactrim DS or Keflex. But if it is not, or should the urinary infection become chronic, you will need a drug that can actually penetrate the inflamed and infected prostate gland. In such cases, I recommend norfloxacin (Noroxin), 400 mg, twice a day for 2 weeks or longer. Men who repeatedly suffer urinary infection because of prostate enlargement should consider prostate surgery.

VARICOSE VEINS: When You Have Eyesores on the Legs!

Have you been keeping your legs and feet covered at the beach or health club because of those unsightly, swollen varicose veins? Is the skin around your ankles brown and maybe even broken with one or more sores that won't heal? If so, chances are more than three to one that you are female. You may not have noticed or paid much attention at the time, but your mother and aunts probably had varicose veins, too. If you're planning to get rid of them, let me warn you up front that no matter what treatment you pursue, those veins are probably going to recur. Yes, even if you have them surgically removed!

Be that as it may, in order to understand how best to deal with varicose veins, you must know why they develop in some people and not in others. Blood flowing within the leg veins must overcome the pull of gravity to get back to the heart and lungs. To help it do so, nature has provided these vessels with valves that function like trap doors. They keep the blood flowing "upward" and prevent it from leaking back down. If you have varicose veins, you were either short-changed in the number of valves with which you were born, or those you do have are not working properly. In either case, the affected veins become engorged with blood that's not flowing as efficiently within them, they dilate, and fluid and pigment seep out through their walls into the surrounding tissues. That's why the legs become discolored, swollen, and, in chronic cases, there is breakdown of the skin and ulcers form.

An important complication of varicose veins, in addition to aching legs and an appearance that precludes your entering a Miss America contest, is *phlebitis* (see page 215). The sluggish blood flow leads to formation of clots on the vein walls, pieces of which may break off, travel to the lungs, and damage them, sometimes with fatal results.

If you have decided to do something about your varicose veins, *don't rush to surgery.* Try conservative measures first. Avoid standing for long periods of time. Support stockings can help move the blood along on its return trip to the heart and prevent it from settling in the legs. I prefer those made by Parke-Davis and Jobst. Wearing these stockings conscientiously will reduce the distention of the veins and the risk of phlebitis. Whenever possible, keep your legs elevated when sitting for any length of time. That's not easy if you're on a long plane flight, given the size of the seats and the amount of leg room in economy class, so get up and move about at least every 30 to 45 minutes (a good reason to sit in an aisle seat). To stimulate the return of blood flow up the veins, rock on your heels whenever you get a chance to do so during the day.

If despite wearing support hose and keeping off your feet as much as possible you continue to have repeated attacks of phlebitis, you need to remove the varicose veins surgically, or have them injected with a chemical that closes them up. Patients often wonder how they can function after these veins have been put out of commission. It's because we have many more of them than we need. Ask anyone who has had a coronary bypass operation in which substantial segments of healthy veins, let alone diseased ones, are removed.

But which should it be, surgery or injection? I prefer the latter because it's just as effective as an operation and much easier to do. Whichever method you choose, chances are your varicosities are likely to recur because faulty valves in the remaining veins will sooner or later result in new varicosities.

Many women, especially those with fair skin, are unhappy about the tiny dilated "*spider veins*" that are sometimes present on the tip of the nose and the cheek bones. They may also discolor the legs, where they don't cause swelling or phlebitis but do occasionally result in chronic discomfort or burning. I do not recommend argon lasers and electrocautery, currently the most popular techniques for the treatment of spider veins, because they can result in scarring. I prefer a new procedure developed at the Massachusetts General Hospital in Boston, called the *pulsed tunable dye laser.* Although it may hurt a little, and sometimes leaves a brownish discoloration that may persist for months, it does work, and is worth looking into if the "spiders" bother you.

VERTIGO (Dizziness): When Your Head Swims and Spins

Vertigo is a miserable disorder that's usually due to some problem in the labyrinths, the structures in the inner ear that are responsible for balance. Here's how an attack usually starts. One morning, after a good night's sleep, without the slightest intimation of any problem, the room suddenly begins to spin and you feel nauseated when you try to get out of bed. You lie down quickly and you're fine as long as you remain on your back. When you try to sit up, you become extremely dizzy again.

The most frequent cause of acute vertigo, especially in previously healthy, young individuals, is a viral infection (viral labyrinthitis). No matter how you treat it, you're going to be "cured" in less than a week. But until you are, there's no use even trying to get out of bed; the dizziness and nausea will overwhelm you. The best drug for the relief of these symptoms is cinnarizine (Stugeron), an antihistamine available in Europe and Canada but not yet approved in the United States. The usual dose is 25 mg, 3 times a day. As with most antihistamines, watch out for drowsiness. Until cinnarizine can be purchased here, use Antivert (meclizine). It comes in 12½ and 25 mg strengths, but try the weaker one first, 3 or 4 times a day. Remember, however, that as with other antihistamines, older men with prostate problems may have difficulty emptying the bladder after its use. If the Antivert alone doesn't help, I often prescribe Transderm Scōp, the antiseasickness patch, or Compazine (prochlorperazine) in a dose of 5 to 10 mg by mouth, 3 or 4 times a day. When you can't keep anything down because of the severe nausea that often accompanies vertigo, use the Compazine in its suppository form—25 mg, twice a day.

When attacks of vertigo come and go every few months year after year, it's not due to viral labyrinthitis. Don't just go on treating the symptoms without a diagnosis. Among the conditions to be considered are a chronic middle ear infection (which antibiotic treatment will eradicate); a disorder of the inner ear involving the labyrinths (more difficult to cure); a tumor pressing on one of the nerves in the area (acoustic neuroma, which can be surgically cured); vascular disease in the brain (that's a tough one to deal with); some cardiac problem (usually a rhythm disturbance); and Menière's disease (discussed below).

Treating Menière's Disease—A Special Kind of Vertigo

In Menière's disease, whose cause remains a mystery, vertigo is accompanied by a loss of hearing and the onset of noises in the affected ear. This association of vertigo, deafness, and tinnitus virtually clinches the Menière's diagnosis. The attacks initially affect one ear, but with time, both are frequently involved. Like all vertigo, it's no fun.

Before diagnosing Menière's, your doctor will first make certain that you are not suffering from low thyroid function, syphilis, or food allergy, all of which can mimic Menière's. Although this disease is usually pretty easy to diagnose with a simple hearing test, an MRI (magnetic resonance imaging) test is sometimes required.

You can reduce the frequency of attacks in Menière's disease by eliminating alcohol, caffeine, and tobacco. The best treatment for the actual symptoms is Antivert, 25 mg, 4 to 6 times a day. As with vertigo due to a viral labyrinthitis, any accompanying nausea and vomiting will respond to Compazine suppositories (25 mg) twice a day.

About 10 percent of all cases of Menière's are so severe and unresponsive to the usual therapy that an operation may become necessary. This should only be an absolutely last resort, since most of the procedures currently performed, though they cure the vertigo, result in permanent hearing loss.

WARTS: They Will Clear Up with Patience

When I was about ten years old, I noticed clusters of warts on my fingers. They weren't painful, but I didn't much like how they looked. When I asked my mother what to do about them, she said, "They're nothing. Leave them alone. Don't pick at them and they'll go away." Mind you, she didn't know that most warts are caused by the human papilloma virus (HPV), but just as she predicted, they always disappeared in several months. Her advice is still good. Leave them alone. They're harmless.

There are several types of viral warts. You can recognize a *common wart,* which appears singly or in clusters on the hands, by its rough surface.

But if it's pigmented black or brown, it may be skin cancer, not a wart, and should be checked out by a dermatologist.

Common warts come and go in many youngsters and teen-agers. They are, to a certain extent, "catching." (Several of my friends as well as my brother and I all had ours at the same time.) So it's probably a good idea, although not crucial, for children with warts to use their own hand towels.

Occasionally a wart that's situated at the edge of the fingernails doesn't go away as my mother said it would! If it becomes painful, it can be removed by "freezing" with *liquid nitrogen.* That's something your doctor can do with a fine spray or a Q-tip. Or, he may recommend electrosurgery or the new carbon dioxide laser treatment, both of which are fancier and more expensive but probably no more effective.

The *plantar wart* is different from the common wart you get on the hands. This one forms on the bottom of the feet, and you are most likely to have picked it up in a public shower. Unlike finger warts, which are usually no more than a nuisance, the plantar variety is painful and often needs to be removed, especially if it is located at a pressure point on which you're standing all day. The liquid nitrogen technique used for the fingers is no good here. You're better off with a Mediplast plaster. But always wash your hands before and after applying it, because if you don't, you can spread the virus to other parts of your body you happen to touch. Nor should you walk barefoot around the house without covering the wart in some way, usually with a plaster. If the Mediplast doesn't work, and the pain is severe enough, the wart may have to be surgically removed.

There's yet another type of viral wart, but this one, condylomata acuminata, or genital wart, is transmitted by sexual contact. It is usually situated at the junction of skin and mucous membranes where most sexual activity occurs—the vagina, the tip of the penis, the anus, and various locations in the mouth. Condylomata have a cauliflower shape when the lesions cluster together. Never treat such genital warts yourself. Local injection with alfa interferon works best. If that's not available, or you find it too expensive, then liquid nitrogen, electrosurgery, or laser therapy will do the job too.

WEIGHT CONTROL: Good Luck!

Here's a secret to which only my wife, my tailor, and all my friends are privy. I'm overweight, and have been for about 15 years. Standing 6′1″ tall, I currently weigh 200 pounds—25 pounds more than I did in 1973, when I was on a crash diet. Back then, at 175 pounds, I looked and felt great, and I was really proud of having accomplished "the impossible." I was, in fact, so euphoric that I presented most of my "oversized" wardrobe (except for the socks and ties) to the Salvation Army, and bought a completely new one—coats, suits, shirts, the works! But alas, over the next couple of years, I gradually and insidiously began to regain whatever I had lost. I couldn't believe it. I blamed it on the scale; I shaved and removed my wrist watch and wedding ring before weighing. Every ounce counts when you're desperate! It wasn't long before I could no longer get into any of my new clothes. I still can't, but this time I haven't given any of them away. Why keep them if they don't fit? Because I am determined to hit 175 pounds again some day—soon, maybe.

My scenario is familiar to the millions of weight yo-yos who go through life constantly and compulsively engaged in losing weight. Here are the statistics. Only one person in seven or eight who goes on a diet maintains any substantial weight loss for more than a year or so. That's because in order to *stay* thin, you must, for the rest of your life, keep track not only of the *quantity* of food you eat, but also its caloric content—ounce for every puny ounce. Mind you, there are those who *can* do it. My wife, who is 5′8″ tall, weighed 145 pounds when we were married. I thought she looked great then, and loved her just the way she was. But she decided that she was too heavy. So she went on a diet that, as all our friends and dinner companions know, she has maintained without deviation for thirty years! She has been a steady 120 pounds (give or take a pound or two during holidays) all that time. Unlike you and me, she has the self-discipline and motivation to do it. At this stage of the game, she no longer even thinks about what she eats or how much. She has a built-in "appestat" that lowers her knife and fork at just the right moment. How can she be satisfied with so little food? "It's easy," she says. "My stomach has shrunk."

What complicates the goal of weight reduction for many people is that there are often other dietary factors to be considered in addition to the number of calories consumed. For example, there are those who must also monitor fat and cholesterol (if they are prone to vascular disease and have abnormally high blood fat—lipid—levels), watch their sugar if they are diabetic, be careful about the quantity of animal organ meats they eat if they have gout, reduce protein in the presence of kidney trouble, alcohol if they suffer from liver disease, salt in the presence of hypertension and heart failure, and dairy products for lactose intolerance. If you need to worry about any of the above and keep track of calories as well, you will be spending an inordinate amount of time just planning menus. Most people don't want to go through life so fixated and tied down; they don't think it's worth the trouble, and neither do I!

But alas, we are not the masters of our dietary destiny, for there is a Mafia of book writers, health faddists, vitamin manufacturers, talk show hosts, friends, and even enemies, all of whom have vested interests in our nutrition, and who simply won't leave us be. Their ultimate objective, of course, is to have us eating like rabbits, turtles, or birds (canaries, that is, not vultures)! And they often succeed, at least for a while. We grimly starve for weeks and months until one day the whole dieting process breaks down when we're overwhelmed by the aroma from a pizza parlor or steak house. Emancipated, we begin to enjoy life again, wallowing in the unbridled enjoyment of eating forbidden foods—until a new guilt merchant comes along. But after a few weeks, it's usually back to the dieting, the daily weights, the thrill of watching the pounds melt, and then the despair that accompanies the inevitable insidious weight increase. Since man is basically an optimistic animal, we rarely throw in the towel (Oprah Winfrey is an exception) because each and every one of us is sure we're going to make it with the next pill or the next diet. It's like buying a lottery ticket, and the chances of winning are just about as great.

So how come there are so many people around whose weight appears to be ideal? How do *they* do it? And why can't you and I? The answer lies in genes—theirs and ours. These individuals are *constitutionally* thin; they don't need to exert any willpower to stay trim. They can eat virtually what and however much they like without gaining any weight.

If you don't have that kind of genetic profile, the odds are stacked against your staying thin because of a basic natural protective mechanism. When an animal is food-deprived in its natural habitat, certain hormonal and metabolic processes come into play that were designed to prevent starvation. When there isn't enough to eat, the body metabolism is lowered so that fewer calories are burned when a given task is performed. This state of affairs continues until the food supply is plentiful once more. Were it not for this energy-saving mechanism, millions of animals would otherwise die during a famine, and entire species have become extinct. Now, nature makes no distinction between a fat, civilized human who deliberately starves in order to lose weight, and a hungry animal who has no choice in the matter. The same protective processes operate for both. *The less you eat, the fewer calories you expend.* As you begin to diet, you lose weight initially, but that doesn't continue for very long as your body shifts to a lower fuel gear. When you stop losing weight, you're apt to become perplexed, frustrated, and angry. Here you are, dieting like crazy, depriving yourself of all the delicious "no-no's" about which you now fantasize, and yet you don't shed another ounce! What's worse, should you deviate even slightly from your rigid diet and sneak in a delicious little "nothing" between meals, you *gain* an inordinate amount of weight.

Disillusioned with the gurus, you visit your "traditional" doctor. If he or she can't provide a quick fix at this critical juncture (which is usually the case), you start hunting anew for a new appetite suppressant or weight-loss program, or you buy the latest best-selling diet book. That's *not* the route to go. I've yet to find an appetite suppressant that maintains its effect for more than a few weeks. Although most are safe if taken *exactly* as directed by fat, *healthy* people, they can sometimes elevate blood pressure and leave the heart more irritable and cause it to beat irregularly, and they often interfere with sleep. They may also mask the fatigue that may be a warning sign of some other underlying process. So *I never prescribe appetite suppressants to anyone.* They are only a fragile, temporary crutch, and are potentially harmful.

There are several commercially available "miracle" weight-reduction diet programs, many of which are franchised by large, for-profit companies. One of the oldest and best known is *Optifast,* developed years ago by Dr. Victor Vertes and his associates in Cleveland. This formula was bought by the Sandoz Company, which now packages and markets it to private doctors and clinics throughout the country. When you join this program, the professionals who run it sell you the Optifast, monitor your compliance and tolerance, and also provide ongoing psychological motivation. You are first checked to make sure your obesity is not due to or accompanied by some other disease, for example, a sluggish thyroid. Optifast operators say they won't enroll anyone who is less than 50 pounds (30 percent) over ideal weight, but I have the feeling that if you turn on the charm, you'll probably be able to find a facility that will bend the rules a little.

The Optifast program is usually continued for about 26 weeks, during which time you're seen by the supervising doctor at least once a week. Your blood is analyzed at regular intervals. During the first 12 weeks you subsist entirely on the powdered diet formula, which provides 800 calories a day. In the remaining 14 weeks, your caloric intake is gradually increased, first by adding some regular food to the Optifast, and finally replacing it completely. Most participants, however, continue to substitute at least one meal a day with the powder from time to time, especially in anticipation of or after a binge.

Many of my patients have tried Optifast. There's no question that it works for several months or more if followed as directed. But in my own practice I have observed that meaningful weight reduction is rarely maintained for longer than a year. On the plus side, however, I have never observed any serious negative consequences from Optifast. At this time, the program costs about $100 a week, which includes both the formula and the professional fees.

There's another diet plan called *Nutri-System,* which sells prepared food that has been freeze-dried, frozen, and calibrated to provide a daily total intake of about 1,000 calories. It is run by nutritionists and psychologists who meet with you at their "center" once a week, all of which will set you back about $50 a session. To the best of my knowledge, there are no doctors involved on an ongoing basis, although I am sure consultants are available when necessary. As with Optifast, normal foods are added to this diet as you go along depending on how much weight you've lost. I have no personal experience with Nutri-System.

Overeaters Anonymous is a voluntary, nonprofit "club," very much like AA, where dietary advice is coupled with motivation. It exploits the "misery likes company" concept, and makes for good fellowship, regardless of how well you do with your weight. Some people can be successfully stimulated by the example of others. It's worth a try.

The bottom line in all these weight loss programs is that fewer than one in five participants ever complete them, let alone continue to be thin for any length of time afterward.

A word of caution. A *very low calorie diet,* less than 800 calories a day, may lead to dehydration, impaired liver function, and cardiac rhythm disorders. There were at least fifty-eight reported deaths from the "special" diets used in the mid 1970s. Six of these were said to have occurred on the Cambridge Diet, which, in its heyday, was the basis of a best-selling book! The rest were associated with liquid protein diets. So, if for any reason you do plan to follow *any* very low calorie regimen, make sure to take extra vitamins and minerals, especially potassium.

The Time Calorie Displacement Diet: A Safe, Effective Diet for Those with Willpower

Every now and then I come across a truly dedicated dieter who, like my wife, can stick with it. If you think you have that potential, here is an approach that can be effective.

In 1983, I came across a book called *Time Calorie Displacement Approach to Weight Control,* by Dr. Roland Weinsier and his colleagues, published by the George Stickley Company in Philadelphia. I'm not sure it's still in print. Unlike other diet books, this one wasn't accompanied by hype or hoopla, and I don't remember seeing it on any best-seller list. It should have been. The basic theory of the authors, with which I agree, is that in order to be effective, a weight-reduction diet should take into account not only caloric content but also *how long it takes to eat a particular food.* Clearly, a dish that's very filling and requires lots of munching and chewing is preferable to one that, in small portions, is very rich but does *not* satisfy hunger. For example, it takes less than 5 minutes to consume 9 ounces of butter, which contain 2,400 calories. You would have to eat 30 pounds of vegetable salad to provide the same number of calories, and that takes hours to do. Dr. Weinsier has grouped foods into categories according to their filling capacity (satiety quotient) and caloric content. This permits you to balance your daily diet in terms of vegetables, fruits, starches, milk, meat, fish, and fat. He has kindly given me permission to reproduce some tables containing this information. Take your time; look at them carefully. If you're serious about losing weight, this may be the way to do it.

In Table I, on the left are *low-calorie, high-bulk, slow-eating foods* like artichokes, asparagus, celery, lettuce, mushrooms, spinach, tomatoes, and so on. By contrast, the *high-calorie, low-bulk, fast-eating fat foods* like ricotta, creamed cheese, nuts, and oils are all the way over on the right. The rest are in between.

TABLE I: **Spectrum of Caloric Density of Various Food Groups (cal/oz)**

Low-Calorie High-Bulk Slow-Eating		**Lean Left**			**High-Calorie Low-Bulk Fast-Eating**
Vegetables 10	**Fruit** 15	**Starches** 50	**Meats** 75	**Sweets** 150	**Fats** 175

Table II indicates that in any 24-hour period you are permitted 4 different dishes from the vegetable group, 5 *from each* of the fruit, starch, and meat sections, 3½ from the fats, ½ serving from the milk group—and you *still won't exceed the 1,000 calories per day* that overweight women should consume, or the 1,200 calories recommended for men. (You may, of course, require more or possibly less, depending on your weight, how many pounds you want to lose, and how quickly, as well as the number of calories, you normally burn in the course of your daily activities.)

TABLE II: **Guide to the Number of Serving Equivalents to Select from Each Food Group Per Day for Weight Reduction**

Diet	Vegetable Group	Fruit Group	Starch Group	Milk Group	Meat Group	Fat Group
A (975 cal)	3	4	4	½	4	3
B (1030 cal)	4	4	4	½	4½	3
C (1200 cal)	4	5	5	½	5	3½
D (1370 cal)	5	6	6	½	5½	3½
E (1540 cal)	5	7	7	½	6	4
F (1700 cal)	5	8	8	½	6½	4½

Table III (on pages 288–92) shows you how to design a meal consisting of items from each category, based on the target number of calories.

Supposing you want to limit your intake to 1,200 calories. You know from Table II that this permits 4 serving equivalents from the vegetable group. So you can, for example, enjoy a cup of mixed salad, a cup of spinach, one small tomato, and a cup of green beans. You're also allowed any 5 servings from the fruit group. A reasonable assortment might include half an apple, half a small banana, a third of a canteloupe, two cherries, *and* half a grapefruit. But there are other combinations from among which to choose, and they may be fresh, frozen, or sugar-free (canned). You are further entitled to 5 servings from the starch group. That might include a 3-inch corn on the cob, *and* ⅔ cup of green peas, *and* a medium baked potato, *and* ½ to ⅓ cup of cooked beans, *and* ½ cup of cereal. From the milk group you're permitted ⅔ cup of low-fat milk or buttermilk, yogurt, or milk powder. And look how generous you can be with meat! Go ahead and wallow in 5 servings, each of which contains about 75 calories. Choose among beef, pork, lamb, shrimp, and skinned poultry. If you are especially passionate about any one of these, you can use all 5 servings for it alone. But then, of course, you may not eat any of the others. Finally, with respect to the fat group, which contains high-calorie, low-bulk, fast-eating fats, oils, and nuts, you may have 3½ servings. So, choose from among a tablespoon of creamed cheese, a teaspoon of regular butter, 1½ teaspoons of peanut butter, 2 teaspoons of regular salad dressing, and unsalted pumpkin, sesame, or sunflower seeds. Why, you can even enjoy some riccota!

This Time Calorie Displacement Diet also permits you on special occasions, twice a week, to splurge on foods that contain about 200 calories per serving: for example, ½ cup (1 scoop) of ice cream, 10 potato chips, ginger snaps, canned fruits, and other delicacies.

What I like about this plan is that you can have virtually any food you like, so that you're less apt to go on a binge because you've been totally deprived for too long of something you really enjoy. It also provides an adequate mineral and vitamin intake without the need for any additional supplements. According to Dr. Weinsier, half the number of individuals who have followed the Time Calorie Displacement approach have maintained their weight for at least two years.

Remember, however, that when following this or any other diet, you should also be *exercising regularly and aerobically* (walking, running, dancing). Also, no diet has any chance of success if you continue to consume all the extra calories in more than token amounts of alcohol in any form—beer, wine, or hard liquor.

TABLE III: **Time-Calorie Displacement Chart**

Low-Calorie High-Bulk Slow-Eating					**"Lean" Left**
Vegetables		**Fruits**		**Starches**	
Eat at LEAST ____ serving equivalents/day	20 cal/ serving equiv.	Eat at LEAST ____ serving equivalents/day	40 cal/ serving equiv.	Eat ____ serving equiv./day	70 cal/ serving equiv.
		Preferred Foods			
Servings listed are for **raw** vegetables: all cooked vegetables equal ½ c		Fresh, frozen, sugar-free		Cooked starches should represent at least ½ of total intake	
				Cooked	
Artichokes	½ bud	Apples	½ med	Beans, cooked	
Asparagus		Apricots	2	Lentils	⅓ c
Bamboo shoots	½ c	Banana	½ sm	Kidney	⅓ c
Bean sprouts	1 c	Blackberries	½ c	Lima	⅓ c
Beets	1 med	Blueberries	½ c	Pinto	⅓ c
Broccoli	1 c	Cantaloupe (5-inch diam.)	⅓	Soy	¼ c
Brussels sprouts		Cherries, red sweet	11 med	White	⅓ c
Cabbage	1 c	Cherries, red sour	⅓ c	Cereal, cooked	
Carrots	1 sm	Grapefruit (4-inch diam.)	½	Buckwheat	½ c
Cauliflower	⅔ c	Grapes		Millet	½ c
Celery (5 inches)	6 stalks	Purple	12	Oatmeal	½ c
Cucumbers	1 lg	Green	20	Ralston	½ c
Eggplant		Honeydew (6-inch diam.)	¼	Seven-grain	½ c
Greens	1 c	Mango	½ sm	Wheatena	½ c
Green beans	1 c	Nectarine	1 med	Barley	½ c
Green peppers	1 lg	Orange (2½ -inch diam.)	1	Bulgar	½ c
Kohlrabi	½ c	Papaya	⅓ med	Corn	
Lettuce	5 lg leaves or ¼ head	Peach	1 med	On cob	3-inch ear
Mushrooms	7 sm	Pear	½ med	Kernels	½ c
Okra		Pineapple	1 sl or ½ c	Peas, blackeyed	⅓ c
Onions (3-inch diam.)	½	Plums	2 med	Peas, green	⅔ c
Radishes	10 sm	Raspberries	½ c	Potato	
Rutabagas	½ c	Strawberries	10 lg or ¾ c	Baked	1 med
Salad, mixed	1 c			Boiled	1 med
Scallions	3			Mashed	½ c
Spinach	1 c			Pumpkin	¾ c
Squash, summer	1 c			Rice, brown	½ c
				Squash, winter	½ c
				Sweet potato	½ med

TABLE III: **Time-Calorie Displacement Chart *(continued)***

"Lean" Left				**High-Calorie Low-Bulk Fast-Eating**	
Dairy		**Meat Group**		**Fats, Oils, Nuts**	
Eat ______ serving equiv./day	80 cal/ serving equiv.	Eat at MOST ______ serving equivalents/day	75 cal/ serving equiv.	Eat at MOST ______ serving equivalents/day	45 cal/ serving equiv.
		Preferred Foods			
May exchange serving for serving with meat group		Weigh portions after cooking			
		Fish			
Skim, nonfat		Bass	1 oz	Avocado (3¼ × 4 inches)	⅛
Buttermilk	1 c	Catfish	1 oz		
Evaporated		Cod	1½ oz	Butter	
diluted		Crabmeat	3 oz	Regular	1 t
1:2	1 c	Clams	10 med	Whipped	1½ t
Milk		Flounder	1¼ oz	Cream	
powder	⅓ c	Haddock	1½ oz	Half & Half	2 T
Milk	1 c	Lobster	2½ oz	Sour	1½ T
Low fat, 1–2%		Oysters (raw)	8 med	Whipping	1 T
Milk	⅔ c	Perch	1½ oz	Non dairy	3 t
Yogurt,		Salmon	1½ oz	Cream cheese	1 T
plain	⅔ c	Scallops	2 oz	Margarine	
Whole		Shrimp	20 or	Diet	1 T
Buttermilk	½ c		2½ oz	Whipped	1½ t
Evaporated		Snapper	1½ oz	Regular	1 t
diluted	½ c	Tuna (water	2 oz	Mayonnaise	
1:2		pack)		Regular	1½ t
Milk	½ c	Poultry		Low calorie	1 T
Yogurt		Chicken (no		Nuts, unsalted	
plain	½ c	skin)	1½ oz	Almonds	7
		Turkey (no		Brazil	2
		skin)	1½ oz	Cashew	4
		Meat		Hickory	7
		Beef	1 oz	Peanuts	9
		Pork	1 oz	Pecans (halves)	5
		Lamb	1 oz	Walnut (halves)	5
		Veal	1 oz	Oil	1 t
				Peanut butter	1½ t

TABLE III: **Time-Calorie Displacement Chart *(continued)***

Low-Calorie High-Bulk Slow-Eating		“Lean” Left			
Vegetables		**Fruits**		**Starches**	
Eat at LEAST ____ serving equivalents/day	20 cal/ serving equiv.	Eat at LEAST ____ serving equivalents/day	40 cal/ serving equiv.	Eat ____ serving equiv./day	70 cal/ serving equiv.
		Preferred Foods			
Tomatoes	1 sm	Tangerine	2 sm	**Dry**	
Turnips	½ c	Watermelon	¾ c	Bread, whole grain	1 sl
Water chestnut	4			Cereal, dry	
Zucchini	1 c			All Bran	⅓ c
				Bran Buds	⅓ c
				Bran Flakes	⅔ c
				Raisin Bran	½ c
				Shredded wheat	½ c
				Most	½ c
				Crackers	
				Rye Crisp	4
				Venus Wafers	4
				Popcorn, plain	1½ c
		Occasional Foods (Up to 2 items/wk from each group)			
Canned vegetable	½ c	Canned fruit†		Angle food cake	
Pickle, sour	1 lg	Applesauce	½ c	3 × 4 × ½ inches	1 sl
Sauerkraut	⅔ c	Fruit cocktail	½ c	Biscuit or	
Tomato juice	3 oz	Mandarin		Muffin	½
V-8 juice	4 oz	oranges	½ c	Bread, white	1 sl
		Dried fruit		Cornbread 2 ×	
		Apricots	2	2 × 1½ inches	1 pc‡
		Dates	1½	Cereal, dry, other,	(see
		Figs	1	non-sugared	label)
		Prunes	1½	Cereal, cooked	
		Raisins	2 T	white	½ c
		Fruit juice		Crackers	
		Apple	⅓ c	Graham	2½ sq
		Cranberry	¼ c	Oyster	20
		Grape	¼ c	Soda	5
		Grapefruit	½ c	Pasta, cooked	½ c
		Orange	⅓ c	Rice, cooked	
		Pineapple	⅓ c	white	½ c
		Prune	¼ c	Roll, dinner	1 sm

TABLE III: **Time-Calorie Displacement Chart *(continued)***

"Lean" Left				High-Calorie Low-Bulk Fast-Eating	
Dairy		**Meat Group**		**Fats, Oils, Nuts**	
Eat ______ serving equiv./day	80 cal/ serving equiv.	Eat at MOST ______ serving equivalents/day	75 cal/ serving equiv.	Eat at MOST ______ serving equivalents/day	45 cal/ serving equiv.
		Preferred Foods			
		Cheese		Salad dressing	
		All hard cheese	⅔ oz	Regular	2 t
		Cottage cheese (noncreamed)	½ c	Low calories	2–4T (see label)
		Ricotta cheese (part skim)	¼ c		
				Seeds, unsalted	
				Pumpkin	1 T
				Sesame	1 T
				Sunflower	1 T
		Occasional Foods (Up to 2 items/wk from each group)			
		Bologna	2 oz‡	Bacon	1 sl
		Cured meats	1 oz	Bacon drippings	1 t
		Duck	¾ oz‡	Chitterlings	¼ c
		Egg	1 med	Cracklings	1 t
		Frankfurters	1⅔ oz§	Gravy	2 T
		Goose	¾ oz‡	Salt pork	¼ oz
		Luncheon meat	1½ oz§	Alcohol, count 2	fats
		Organ meat	1 oz	Beer	
		Ricotta cheese	2½ T	Regular	3 oz
		Sausage	1¼ oz§	Lite	5 oz
				Liquor	¾ oz
				Wine	
				Dry	2 oz
				Sweet	1 oz
				Lite	4 oz

†Unsweetened.
‡Count 1 fat.
§Count 2 fats.

TABLE III: **Time-Calorie Displacement Chart** ***(continued)***

Special Occasion Foods
(Up to 200 cal/wk)
Calorie Content of Selected Items

Item	Calories
Cakes	
Pound (3 × 3 × ½ inches)	125
Brownies (2 × 3 × 2 inches)	145
Cake with icing (2 × 3 × 2 inches)	210
Candy—1 oz	150
Cereals, dry, sugared (see label)	
Cookies	70
Ginger snaps	20
Vanilla wafers	30
Crackers, snack (see label)	
Cranberry sauce 1 T	30
Donuts	
Plain 1 med	125
Jelly 1 med	225
French fries 10 pcs	135
Fruits, canned, sweetened, 2 halves or ½ c	80
Honey 2 t	40
Ice cream ½ c, 1 scoop	150
Ice milk ½ c, 1 scoop	100
Jello ½ c	65
Jelly 1 level T	50
Juice, sweetened ("drinks") ½ c	50
Pies ⅙ pie	380
Potato chips 10	115
Puddings	
Whole milk ½ c	130
Skim milk ½ c	70
Sherbet ½ c	130
Soda 10 oz	130
Soups, cream 8 oz	150
Sugar 1 t	20
Yogurt, sweetened or fruited 1 c	250

WHOOPING COUGH (Pertussis): To Hear It Is to Fear It

If you have ever been in contact with someone who has whooping cough, or can remember the "whoop" when you had it yourself, you can appreciate how it got its name. Whooping cough is, for the most part, a childhood illness. More than 50 percent of cases occur before the age of two. The vaccine is effective, but regrettably its safety was questioned after a few bad reactions were reported, and over the years many mothers refused to allow their children to have it. As a result, substantial numbers of people in the United States remain vulnerable to this infection because they have had neither the disease (which confers lifelong immunity) nor the vaccination. If your child is under seven years of age, *was* vaccinated, and has subsequently been exposed to whooping cough, arrange for a booster shot to increase the antibody level. But those older than seven who were vaccinated and now exposed, should *not* be revaccinated because of the potential adverse reaction from the vaccine. Unlike some other childhood diseases (like German measles), where we actually want our kids to catch it and be done with it, you're better off without whooping cough, and should obtain your protection from the vaccine. Anyone with this infection should be strictly isolated for at least 4 weeks.

Whooping cough is almost unique among the childhood infections in that it is due *not* to a virus, but to a bacterium. The first symptoms are a bad cough, stuffy nose, and what generally appears to be a "serious" cold. But the next stage presents with exhausting bursts of coughing that continue for what seems like an eternity. The characteristic whoop as the patient breathes in, makes the diagnosis easy. When the whoop has gone, you're on the mend even if a "normal" cough does linger for a while.

Infants under the age of one year with whooping cough should always be hospitalized because of the high incidence of life-threatening complications. But this infection should be taken very seriously at any age, and its symptoms managed with great care. I recommend using the most potent cough medicine there is—one with plenty of codeine—and when the patient is very sick, steroid injections may be necessary.

What should you do if you're an adult vulnerable to whooping cough and have been exposed to it? The incubation period—that is, the interval from when you were infected to when you become sick—is about 7 to 10 days. It's now too late for either the vaccine or immune gamma globulin. Regardless of whether or not you have any symptoms, take Ilosone, a particular form of erythromycin, for a full 14 days—no more, no less. I rarely prescribe Ilosone for any other condition because it has many more side effects than other erythromycin preparations, and often causes severe abdominal pain and sometimes even bloody stools. But it's so effective against whooping cough that it's worth risking the possible toxicity.

The most serious complication of whooping cough is pneumonia. Depending on the organism that's isolated from the sputum, an appropriate antibiotic—again, usually Ilosone or ampicillin—should be taken. Other consequences of this disease are middle ear infections and chronic lung disorders, as well as hemorrhages and hernias from the cough. Some children also develop seizures at the height of the whoop, for which the best treatment is phenobarbital.

WORMS AND PARASITES: Ugh!

There are hookworms and pinworms, fishworms and beefworms; there are intestinal flukes and liver flukes; there are amoebae, schistosomes, and *Giardia*.

I wasn't sure whether or not to include something about worms in this book because there are so many of them, and the treatment is so varied. But then within two months, three or four patients came to the office who were all infected with the same parasite. I had such success treating them, I thought I'd pass the following information along to you.

If you travel widely or eat out often enough, even in your own neighborhood, or consume enough sushi—sometime, someplace, one of those ugly beasties is going to take up residence somewhere in your interior. There is no way to stop a worm from wriggling its way into your life, usually causing some kind of bowel disturbance. Only scrupulous attention to what you eat, and not medication, can protect you. But don't confuse infestation by a *parasite* with "turista" or some other *bacterial* infection that you *can* prevent with drugs.

When your friends, your travel agent, or your doctor tell you to be very careful about what you eat and drink while traveling abroad, particularly in Asia, Africa, or South America, don't dismiss it as small talk. So many people have the macho attitude that *their* gut and *their* resistance are strong enough to fend off any food- or water-borne infection. Let me give you an example of how wrong they are. My wife, a few close friends, and I recently sailed on a cruise boat from Singapore to Bangkok. I had cautioned all the members of our party against drinking *any* tap water, not even the small amount needed to swallow pills. They were to use only bottled water, preferably sparkling. After 11 days on board ship we checked into a first-class hotel in Bangkok. We were very careful not to eat any unpeeled fruit, we avoided ice in our drinks, and we restricted our liquids to alcohol and brand name bottled water. However, one of our number, in what seemed like a trivial lapse of discipline, took a few sips of water from her bathroom tap in order to swallow her daily vitamins. Three or four days later, she felt a little queasy, and noted some discomfort in her upper abdomen that she thought was from overdoing the food and wine. These symptoms persisted, worsened a bit, and then the diarrhea began. By the time she arrived back in the United States a few days later, she had lost seven pounds, and continued to suffer several bouts of diarrhea each day. The pain in her upper abdomen was bad, too, and her doctor, suspecting an ulcer, gave her ranitidine (Zantac) and antacids, advised frequent small feedings, and forbade any alcohol, caffeine, tobacco, or aspirin. This regimen helped a little, but not much. The working diagnosis was peptic ulcer because of the pain in the upper belly; the diarrhea was attributed to all the antacids she was taking. After 7 or 8 days of treatment, her doctor decided to have her stool analyzed for parasites, just in case. To his and everyone's surprise, it was found to be teeming with *Giardia*.

I've gone into all this detail because this particular parasite is found *everywhere,* not just abroad. You don't have to go to Bangkok to get it; it is the most frequently diagnosed parasitic infection in the United States. It's usually acquired from tap water, but you can also pick it up from carriers like infected food handlers, who themselves are without symptoms. So if you have diarrhea, lots of gas and bloating, have lost weight, and are tired, the first thing to do, especially if you've been on a trip, is to get a stool test for bacteria and parasites.

I was very critical of the way the doctor handled the case I described. Specifically, I thought he waited too long before sending the stool for analysis. I couldn't really be angry with him, however, because there were extenuating circumstances. You see, the patient was my wife, and *I* was the doctor!

So the bad news about *Giardia* is that it's everywhere, and it doesn't take much for this miserable critter to make you sick. The happy news is that treatment is easy and very effective. Metronidazole (Flagyl), 3 tablets a day, 250 mg each, for 7 to 10 days will usually eradicate the parasite and clear up all your symptoms. But here's one of those bizarre quirks that we so often encounter in this country. If you look in any textbook of medicine for the treatment of giardiasis, or if you call the Food and Drug Administration and ask them about it, you will be told to take quinacrine. Not a word about Flagyl! Quinacrine *is* effective, but it causes lots of side effects—gastric irritation, dizziness, yellow skin and eyes, behavioral changes, and skin rashes. In my experience, both personal and professional, I've found Flagyl much better against *Giardia,* although it too can produce nausea, headache, a dry mouth, and a metallic taste. I'm legally obliged to tell you, however, that Flagyl is considered "investigational" by the FDA for that purpose. By the way, it cured my wife in less than three days. Remember that you *must abstain totally* from all alcoholic beverages while taking this drug because it has an Antabuse-like effect. (See page 7.) If you have a drink or two during the course of therapy, you will regret it.

There's another parasite called *Blastocystis hominis,* frequently found in stool cultures of "normal" humans. It does not usually cause any symptoms, so that when it is present together with another organism, you should treat the other one first. However, if you have diarrhea, and the *only* bug that shows up in stool analysis is *Blastocystis* in large numbers, it should be treated—again, with Flagyl.

The most common intestinal parasite present in Americans who have returned from Asia or Africa is *Schistosoma.* If it is identified in your stool, ask your doctor for praziquantel (Biltricide). It will eliminate the infection in one day.

A final word about parasites. After any treatment for whatever parasite is found in your stool, always send another specimen for examination a month or 2 after your symptoms have disappeared to make sure that you are not a carrier. This is especially important if you handle food, not only in restaurants, but in your own home. The parasites may no longer bother *you,* but you may continue to be a source of infection for others.

And finally, the best treatment for long-term good health:

EXERCISE: Surprise! More Is *Not* Better

Many of today's medical "facts" are destined to become tomorrow's myths. For example, in the relatively short span of my own career, there has been a complete turnabout in the approach to several critical health problems. Take high blood pressure, a major killer and crippler. As a medical student, I was taught that as long as an elevated pressure wasn't causing any symptoms, it was best left untreated. (Of course, the fact that at the time there was virtually nothing available with which to reduce it effectively may have had something to do with that opinion.) Today, almost everyone knows how important it is to control hypertension. Not so long ago an elevated cholesterol level was not understood to be a health hazard. Today, you don't dare eat an egg in public without risking hostile stares. There are many other areas in which there has been a complete reversal of opinion as a result of new knowledge; one of these is *exercise*. In the early years of my practice, every heart attack patient remained in hospital *in bed* and at *complete rest* for a full six weeks even if there were no complications. With time and experience it became evident that such conservatism was not only of no benefit, but was, in fact, harmful. It left patients feeling weak, without muscle tone, their bones thinned because of calcium loss, and with a vulnerability to dangerous blood clots to the lungs. As a result of these observations, we've been getting patients out of bed earlier and earlier after their myocardial infarction. These days, someone who has had a heart attack is virtually up and about soon after the ambulance reaches the hospital!

After prolonged bed rest was shown to be dangerous, it seemed logical to conclude that *deliberate physical activity* was beneficial. Unfortunately, the pendulum swung completely in the opposite direction. And so until fairly recently, regular, rigorous workouts, almost to the point of pain and suffering, were the "in" thing. Any level of activity short of that which resulted in exhaustion was believed to be a waste of time or effort. This gung-ho approach deterred a great many people from performing moderate exercise on any regular basis because they felt it was a waste of time. This was especially true for the elderly and those with a previous cardiac history.

Happily, reason and moderation prevailed after 1984, when the U.S. government convened a panel of twenty experts to review all the available evidence relating to the potential benefit of exercise. They concluded that a *tolerable* amount of planned physical activity can have an important positive impact on health and the quality of life. In other words, exercise that is actually enjoyable is also healthy and beneficial. (Perhaps that's what President Bush meant when he referred to a "kinder, gentler America.") I expect that as a result of these findings, there will be fewer of those gasping, exhausted, agonized joggers going by your house every day in their quest for fitness.

Unfortunately, despite this latest recommendation, 40 percent of us continue to be almost completely sedentary. On our way to work in the morning, we walk a couple of blocks to the subway in the city, or to the bus station in rural areas, sit for an hour, then saunter another 100 yards to the office. During the day, if we work in a high-rise building, we avoid the stairs like a plague, even for just one or two flights—and prefer to wait 5 minutes or more for an elevator. In the evening we retrace our steps home, we eat dinner, watch television, and go to bed. (Sex doesn't count.)

If you're a woman, that kind of life-style will end up giving you osteoporosis; male or female, you'll have more trouble keeping your weight and blood pressure normal (the incidence of hypertension is 35 to 50 percent higher among physically inactive persons); you'll be a better candidate for developing diabetes in middle age; with less strength in your muscles, and diminished flexibility of your joints, you'll be moving around like an oldster before your time; you'll be more likely to need tranquilizers during a crisis (there's nothing better for mental health than regular exercise); and the risk of your coronary arteries becoming blocked will increase, too. By contrast, persons who exercise regularly have a lower death rate, not only from heart disease, but from all causes. Physical fitness becomes more important with age, at least up to 75 years, so you shouldn't reduce your exercise level as you get older. You can't rest on the laurels of physical prowess in your younger days; it's not how active you *used* to be that counts, but what you are doing *now*.

Exercise for Those Who Are Still Healthy and Want to Stay That Way

As far as I'm concerned, the two essential ingredients of any exercise regimen are safety and enjoyment. You're not likely to spend precious time and effort on something you hate to do. Assuming that you are in good health, you do have a wide choice of *pleasurable* activities from which you can benefit.

How much exercise do you need? Here are the specifics: Burning as little as 500 kcals *per week,* especially in some form of weight-bearing exercise like walking, will measurably decrease your risk for osteoporosis, high blood pressure, heart disease, diabetes, and obesity. You can reduce that risk even further by raising the level of exercise to 400 kcals *per day.* At this point you may well be wondering what kcals are, and how you can expend 400 or 500 of them. Kcals are an index of energy. Walking at a comfortable pace for 30 minutes burns up about 230 kcals. Do that only three times a week, and you can easily satisfy the minimum objective of 500 kcals. Biking for 30 minutes at a leisurely 5½ miles an hour will use up about 200 kcals; if you mow your lawn for 30 minutes, you will expend 300 kcals; weeding the garden costs 200 kcals; playing an enjoyable golf game for only 30 minutes will wear off 250 kcals. If you like to swim (the most popular form of physical activity in the United States today; walking is a close second), a slow crawl for 30 minutes will get rid of approximately 400 kcals. And you needn't "crawl fast" either, unless you want to, because doing so burns off only 50 more kcals in the same 30 minutes. Exercising for periods longer than 30 minutes at a stretch is not really of any greater benefit. *Vigorous exercise* may leave you more fit and provide your muscles with greater tone and strength, but strictly in terms of risk reduction, *500 kcals a day* is all you need.

It's never too late to begin exercising regularly, even if you've been inactive all your life. In fact, persons who have just sat around for years reap the richest health benefits after they start any one of the very low levels of exercise described above. If you fall into that category, go for a 30-minute walk at a comfortable pace three times a week. Then make it a daily outing for the same 30 minutes, but walk more briskly. As far as I'm concerned, that's enough exercise for most people. But you may do more if you enjoy it. Once they get into the habit of walking, most people become "addicted."

Exercise for the Cardiac Patient

What I've described above applies to individuals who are basically healthy without any previous history or symptoms of heart disease. But exercise is also desirable for most cardiac patients *provided it is done properly, at the right time, in the right amount, and only after careful cardiac evaluation.*

Exercise improves cardiovascular fitness; it reduces the cardiac rate at rest so that the heart works less strenuously and more efficiently. During exercise, the heart rate rises less steeply in individuals who are fit, and so results in greater endurance, as well as less shortness of breath or chest discomfort. Exercise is also the best way to raise the HDL ("good") cholesterol level and to reduce LDL ("bad") cholesterol.

An additional benefit of planned exercise in measured amounts in someone with a heart condition is that it disabuses them of the notion that the heart is fragile, and that their illness has "crippled" them for life. The heart is actually one of the toughest organs in the body, and recovers very nicely from most insults. Unfortunately, the mind doesn't always follow suit, and too many patients, men and women alike, remain psychologically crippled after a heart attack. So, participating in an exercise program is "proof" that you are not "terminal" or a "basket case."

If you have any type of heart condition, before beginning a physical fitness program you should always consult a doctor or a cardiologist to determine if exercise will be safe for you, and if so what type and how much.

You should not be exercising if you have *severe* heart failure and are short of breath at rest, if you suffer from certain disturbances of cardiac rhythm, experience angina at rest or during the night, or if you have a valvular disorder called aortic stenosis.

Walking, biking, and dancing are especially good forms of aerobic exercise for cardiac patients because they make use of large muscle groups. Do your walking outdoors if the weather is not too hot, cold, or humid, or on a treadmill indoors. If you live near a shopping center, take advantage of the indoor mall; most of them are air-conditioned in the summer and heated in the winter. But make sure that whatever activity you choose is repetitive and rhythmic. Avoid sports like tennis, for example, and especially singles, that require sudden bursts of activity. Nor do I recommend isometric exercises, particularly those involving the arms—sawing wood, shoveling, or lifting weights. Don't carry anything heavier than 20 pounds going upstairs or 60 pounds on the level, even if it means the embarrassment of having your wife lugging the suitcases while you walk along blithely unencumbered.

Wait about 1½ hours after eating before you start any physical activity. While you're digesting your food, the heart is pumping extra blood to the stomach. Exercising at the same time imposes the additional burden of simultaneously supplying your muscles *and* your gut.

I usually discourage swimming for 3 or 4 months after a heart attack because some patients complain of more angina when they swim during that interval. And you wouldn't want to develop chest pain in the middle of an Olympic-sized pool.

I'm often asked by men and women recovering from a heart attack how soon they may safely resume sexual activity. My advice is to wait until you can *comfortably* walk up and down two flights of stairs without unusual shortness of breath. And I am not being facetious when I also caution that such sexual activity not involve extramarital partners or new and exciting ones because of the additional emotional stress, fear, guilt, or even exhilaration potentially associated with such pursuits.

No matter what exercise regimen you choose, you'll know that you've done more than you should if it leaves you unusually tired or short of breath, or if it causes nausea, vomiting, or insomnia. Next time, stop short of that exercise level.

In summary, most "cardiacs" *do* lead normal, healthy lives, but before engaging in physical activity, they should have a careful physical evaluation. Don't accept the "take it easy" answer when you ask your doctor for specifics. Your exercise program must be calculated and prescribed by your doctor, based on data obtained during monitored stress tests, as carefully with respect to amount, type, frequency, and duration as is the dosage of any drug.

Appendix: The Home Medicine Cabinet

HOME MEDICAL SUPPLIES: What You May Need to Be Prepared

Here is a *basic* inventory of medications and medical supplies which I recommend you have available at home for those problems that may require immediate treatment especially during the night or on holidays when most pharmacies are closed.

The first rule in stocking your medicine chest is always to keep an ample supply of those drugs used regularly by anyone in *your* household—for example, nitroglycerin for angina, insulin for diabetes, bronchodilators for asthma, agents to reduce high blood pressure, and anti-ulcer drugs. These are not part of the usual pharmaceutical inventory for the average home, but should always be on hand to tide you over a long holiday weekend in the event that your supply runs out.

In addition to medication for specific ailments that affect you or your family, every home medicine cabinet should contain:

• *Two thermometers,* one oral, the other rectal (the latter for small children and debilitated or older individuals). Although temperatures taken rectally are more accurate than oral ones (they're not affected by hot or cold food and drink), a mouth thermometer is good enough for routine use. But if your eyesight is poor, consider buying one of the digital battery-powered units, which are much easier to read.

• A cold-mist humidifier for treating respiratory symptoms due to colds and the flu.

• A *liquid* antacid for relief of "gastritis" resulting from dietary or alcoholic abuse. I prefer Maalox, Mylanta, or Gaviscon, but there are many other equally good products, none of which require a prescription.

• For "simple" runs or *diarrhea,* have a bottle of Imodium on hand. You can buy it over the counter. Take 1 or 2 tablets every few hours as necessary. If symptoms persist for longer than that, see your doctor. Children who find it easier to take liquid medication may be given Kaopectate. The 8-ounce size is convenient.

• It's a good idea to have a *decongestant* in the house for when you have a cold. Buy some Sudafed over the counter in the 30 mg strength.

• For *allergic reactions,* ask your doctor for an antihistamine prescription. I prefer Seldane because it doesn't leave you as sleepy as some of the others. Keep about 20 of the 60 mg tablets on hand, and take 2 a day to control symptoms of allergy.

• For occasional *insomnia,* obtain a dozen or so 25 mg tablets of Benadryl (except if you are an elderly male with prostate problems or if you have glaucoma). If that's not enough, have your doctor prescribe 10 Dalmane capsules, 15 mg strength. Take 1 or 2 at bedtime, but *only over the short term.*

• For *pain control* and to reduce *fever,* you should have both acetaminophen (Tylenol) and aspirin at home (but never give the latter to children with fever due to a viral illness).

• Keep some ibuprofen (Advil, Nuprin, Medeprin) on hand for mild pain—200 mg strength. More severe distress may require ¼ grain tablets of codeine.

• It's important to have *syrup of ipecac* available to induce vomiting when a toxic substance has been swallowed (see page 221). But be sure to call either your local poison control center or doctor before giving or taking the ipecac because there are specific circumstances in which it should not be used. For poisoning and other emergencies, you should have by your telephone and pasted in the medicine cabinet the following phone numbers legibly written: police, fire, ambulance, your doctor, a reliable neighbor—and a reminder to use 911 when time is of the essence.

• I advise my patients to have some *antibiotics* available, even though I insist they consult me before using them. There are times when they need an antibiotic and I would prefer they not wait until the pharmacy opens in the morning. But store only a very small amount, just enough to get you started, because these drugs lose their potency after they have expired, and you may not have occasion to use them for months or years. I suggest 12 tablets of oral penicillin (Pen VK 500 mg if you are not allergic to it), 12 doxycycline (Vibramycin), 100 mg strength, and some Bactrim DS (if you are prone to urinary tract infections and are not allergic to sulfa)—to be taken in whatever dosage your doctor prescribes. *Never self-medicate antibiotics.*

• To help *clean wounds* I like Betadine cream. You should also have available an antibiotic ointment or cream like Neosporin.

• Everyone gets *splinters* now and then, which can be removed by Mediplast plasters, so have a few of them on hand. You'll also need some *tweezers* and a 16-ounce bottle of 70 percent isopropyl alcohol (rubbing). The latter is not only a good antiseptic, it makes you feel wonderful when rubbed into aching muscles.

• For *bites,* stings, mild dermatitis, or allergy, you will want some hydrocortisone ointment or cream. There are several over-the-counter preparations available (like Cortaid) that are effective for anything from a sunburn to poison ivy.

• Everyone needs a *heating pad* now and then, as well as an *ice pack,* 6 to 9 inches in diameter. Use the cold pack in the first 24 to 48 hours after an acute muscle injury, and the heat thereafter.

• For *nausea,* I tell my patients to keep some Compazine, 10 mg tablets, in stock. If the nausea is very severe and you can't retain the tablet, then a 25 mg Compazine suppository may be necessary. Ask your doctor for a prescription. Many doctors prefer Tigan by mouth or suppository.

• In the event that a break in the skin requires first aid treatment, you should have a box of Band-Aids containing various size patches, some sterile gauze pads (2″ × 2″), waterproof adhesive tape (a roll of 1½″ × 5 yards), scissors, and a 2-to-3-inch wide elastic bandage.

• Every home should have *cough medicine* handy. Ask your doctor for a prescription for a 4-ounce bottle of Actifed-C, and take 1 or 2 teaspoons every 3 or 4 hours as necessary.

Remember, no matter what medications you store, they should all be *locked* away, or otherwise inaccessible to small children. And be sure you store them in a proper environment. For example, if you have angina, you should always keep a dozen or so nitroglycerin tablets on your person, and leave the rest in the original bottle in the refrigerator in order to preserve their potency. Insulin, too, will spoil if left at room temperature for more than a few days.

Finally, always check the expiration date on the container of medicine you buy, or that you have had around for a while. Expired drugs not only lose their effectiveness, but can, in some cases, actually cause you harm—and that even includes aspirin and Tylenol.

GENERIC VERSUS BRAND NAMES: Is There Really a Difference?

In New York State, at the bottom of every prescription blank there is a little box in which your doctor has the option of writing "DAW"—"Dispense As Written." Unless he does so, your druggist is *obliged* to sell you the "generic" preparation of the medicine prescribed for you. Many other states and countries have similar formats, permitting you and your doctor to choose between "brand name" or "generic" medication. Many patients are confused by this distinction, and are not sure which to select. Following is the difference between them.

In the United States, a pharmaceutical company introducing a new drug is protected against "copy cats" by a seventeen-year patent, during which time no one else may sell an identical preparation. Such protection is very important, because if every competitor had the immediate right to reproduce a medication that had been years in development at a cost of millions of dollars, commercial research incentive would be stifled. Mind you, although I fully understand and appreciate the profit motive, I'm nevertheless often dismayed by the exorbitant price tags on some of these "protected" agents, especially when they are unique and life-saving, and the consumer has no choice but to pay what the traffic will bear.

After the patent has expired, the product becomes fair game, and if it is a good one, pharmaceutical companies whose sole function it is to duplicate drugs, not discover them, get into the act. They may then make their own product, identical to the original, but are not allowed to use the same "brand" name. Theirs is called the "generic" form.

The Food and Drug Administration (FDA) in the United States (other countries have equivalent agencies) monitors the quality, safety, and efficacy of these "me too" products in order to protect the consumer. Every generic drug must, by law, possess the same biological activity and/or potency as the corresponding brand name. Aside from the occasional scandal, that is something you can pretty much depend on. The extent to which variations in biologic potency do occur is generally not important.

That being the case, why pay two and three times more for the brand name when the generic form is guaranteed by law to be virtually identical? The truth is that there may be other variations among preparations with equivalent *biological* activity that can make a difference to you: the coating on the tablet; the presence of "fillers" like glucose or lactose in one and not the other; the coloring used; taste "maskers" in the case of a liquid; and so forth. Regulatory agencies pay little or no attention to these "cosmetic aspects" of the manufacturing process because they are biologically inert. But these variants can and do affect the rate at which the preparation dissolves in the stomach, its transit time through the gut, and can even cause side effects that have nothing to do with the "activity" of the medication itself.

Take lactose, a common ingredient in many pharmaceutical preparations. There are millions of people who are "lactose-intolerant" and who develop cramps, gas, and bloating whenever they consume this sugar. The brand name may not contain enough lactose to cause such symptoms, whereas the generic one does—or, of course, vice versa.

Here's another important fact to bear in mind. You may not receive the same generic product each time your prescription is refilled because pharmacists shop around for the best buy they can get in a generic formulation. So you may end up with manufacturer A's product one day, and maker B's the next time. Unlike the brand name, generic products not only fluctuate in price, they may vary in other ways among different makers. That is not the case with the brand name, which is always the same no matter when and where you buy it. Now while it is true that you will not be able to distinguish variations in the manufacturing process in the great majority of drugs, even small differences can be important in anti-epilepsy medication, birth control pills, and in certain cardiac agents.

Given the above pros and cons, what route should you follow when filling a prescription—generic or brand? These days you may not always have a choice. Many HMO's, the Medicaid program, and several third-party insurance payers will only reimburse you for generics. Their policy is that if you want the "luxury" of a brand name, you should pay for it. Where you do have an option, this is what I recommend:

- When there is no significant price differential, buy the brand name.
- When it's more costly (as it usually is), discuss *each medication* with your doctor. In some cases there is no real difference, in which event there's no point throwing away good money on the trade name product. In others, the doctor may insist on the brand name regardless of higher cost.

- If you've been using a generic right along, try to buy the refills from the same pharmacy. You're more likely to get the same manufacturer's product than if you go somewhere else.
- Ask your druggist, each time you renew your prescription, whether the medication comes from the same maker as it did in the last prescription filled.
- Whenever you refill a generic drug and you notice a change in your symptoms, let your doctor and your pharmacist know. For example, if you're having more angina or your epileptic seizures are more frequent, or your heart rate is more irregular, slower or faster than it used to be, the cause may be due to a different brand of nitroglycerin, anticonvulsant, or dilantin preparation, respectively, and not necessarily a worsening of your condition.

In summary, despite recent scams involving a handful of dishonest manufacturers, generic drugs are biologically equivalent to the brand name product and much cheaper. I usually prescribe generics for my patients except in those conditions such as seizure disorders or cardiac disease in which even small variations in potency and method of manufacture can have important clinical consequences.

PAIN KILLERS: On a Scale of 1 to 10

Believe it or not, a pain killer—known technically as an analgesic—is not always the best thing when you hurt. For example, aspirin is useless—and possibly even harmful—when your belly aches from an ulcer. To obtain relief, you need something to neutralize the excess acid in your stomach, or to prevent its formation. When your chest is gripped in the vise of an angina attack, forget about traditional pain relievers and just slip a tiny nitroglycerin tablet under the tongue. It will dilate the narrowed, diseased coronary arteries in seconds, increase the blood flow to the heart muscle, and make your pain disappear.

But there are times when one specifically *does* require an analgesic—after an injury or an operation, or for some chronic, painful condition—anything from cancer to arthritis. In order to choose the right agent, it helps to know something about how and why one feels pain, and the ways in which various medications can relieve it.

There are two components to pain perception; one operates at the actual site where the hurt is felt, the other is in the brain. When an arthritic joint is red, hot, and swollen, or a bone is broken, or body tissue is bruised, a substance called *prostaglandin* is released in the area. There are several different prostaglandins, all of which trigger *local* nerve endings to transmit a "pain" signal to the brain. That's where the second component comes in. The brain interprets that signal and tells you that you hurt in the injured area. Once alerted to the fact that something is wrong, you can take the necessary steps to deal with the problem. For example, if you've broken a leg, you'll keep off it until it's fixed—and if you don't, you will soon be reminded by more pain.

So you have two options in controlling pain sensation: you can either take a medication that blocks or reduces the production of prostaglandins so that fewer signals leave the injured area, or you can use a drug that numbs the brain. In the former instance, the brain doesn't get the message, so you don't know you "hurt." In the second, the signals do reach the brain, but just bounce off it and so you "feel no pain."

Now let's get down to specifics. Aspirin, which is what most people reach for first when they're uncomfortable or aching somewhere, is the classic *antiprostaglandin* drug. The newer nonsteroidal (so called because they do not contain cortisone) anti-inflammatory drugs (NSAIDs) are too, and I have found them to be more potent and in some respects safer than aspirin. By contrast, *narcotics,* the most effective of all analgesics, and acetaminophen (Tylenol) act *directly on the brain* and interfere with its ability to perceive pain. There are several different narcotics. All of the "natural" ones come from the poppy seed, but there are now many synthetic derivatives as well. Generally speaking, you should use an antiprostaglandin drug when the pain is mild and bearable, and a narcotic when it is severe or when the antiprostaglandins don't work. There's no point in taking morphine when a couple of aspirin will do the job.

In addition to considering the *severity* of pain, there's also the matter of how well *you* tolerate it. Some people have very high thresholds for pain, while others cringe when the skin is being punctured for a flu shot or a blood test. But no matter how stoic you are, there will be times when a couple of aspirins aren't enough, like when you've broken your leg or had major surgery.

Your choice of an analgesic should also depend on whether the pain is *acute* or *chronic*. In other words, is it apt to continue indefinitely, as in chronic arthritis, or is it only a transient interlude in your life—like the first few days after an operation? You should have as much morphine as you need to feel comfortable after surgery, but think long and hard about taking it for a bad back that has been hurting you for years, and is likely to continue to do so in the future. If you come to rely on a powerful narcotic to relieve the pain that recurs day after day, week after week—*indefinitely*—you will almost certainly end up addicted.

Given these general guidelines, here is some specific information about various analgesics to help you choose the right one.

Non-narcotic Analgesics

The prototypes of the two mildest pain killers are *aspirin* and *acetaminophen* (better known as Tylenol). They do not require a prescription unless combined with a narcotic such as codeine, which is often the case. Neither of these drugs alone is usually potent enough to end big league suffering, but they will relieve "minor aches and pains."

How many times have you heard it said, "Have a headache? Take some aspirin or Tylenol"—as if they were one and the same and interchangeable. They are not! With respect to how it relieves pain, aspirin is an antiprostaglandin, which acts locally, while Tylenol works directly on the brain as if it were a narcotic, which it is not. Antiprostaglandins, in addition to minimizing pain, also decrease inflammation. Acetaminophen does not. So when your arthritic joint is red, swollen, hot, and hurts, aspirin and the related NSAIDs will help you more than acetaminophen will. However, for other causes of discomfort—a mild injury or a sprained muscle in which there is not a great deal of inflammation—acetaminophen is as effective as aspirin and often preferable.

Here are some other pros and cons of these two drugs:

- Acetaminophen should not be taken by a "drinker" or anyone with liver or kidney disease. It can injure these organs, especially when used regularly and in large doses for several consecutive days or weeks.
- Aspirin, even in small amounts, can cause bleeding from the upper gastrointestinal tract, especially in older people. Acetaminophen does not.
- Aspirin can induce acute asthmatics attacks. Acetaminophen does not.
- Long-term aspirin therapy helps prevent strokes and heart attacks because of its anticoagulant effects; acetaminophen will not.

- Acetaminophen does not have aspirin's effect on the blood-clotting mechanism, so it should be used in preference to aspirin if you have any blood problems, or if you tend to bruise or bleed easily.

Both aspirin and acetaminophen come in various strengths and in different formulations—drops, liquids, tablets, capsules, and suppositories. Choose the format most appropriate for your needs. For example, a liquid is easier to give to children, if it has an acceptable taste. Whenever possible, aspirin should be taken *after* meals, because it is less likely to irritate the gastric lining when mixed with food. Some doctors don't think that makes any difference; I do. But you may take acetaminophen any time—before, during, or after meals.

Should you buy the generic or the brand name products of aspirin and acetaminophen? In my experience, it doesn't make one iota of difference.

Is aspirin *buffered* with an antacid (Bufferin, Ascriptin) easier on the stomach than the plain preparation? I don't think there is enough antacid in these tablets to make any real difference. But *coated aspirin* (Ecotrin) *does* cause less gastric discomfort, because most of the active ingredient is not released until the tablet has left the stomach and is further along in the gut. Even so, *never take aspirin in* any *form if you have an active ulcer.*

NSAIDs versus Aspirin

I believe the NSAIDs are more effective than aspirin for pain control, but they can hurt the liver and kidney. If you're taking them on a regular basis, have your blood tested for liver and kidney function every 3 or 4 months. The NSAIDs also cause salt retention, which is undesirable if you have either salt-sensitive high blood pressure or a cardiac problem, so avoid them if you do. Even though the NSAIDs are less irritating to the gut than is aspirin, they should not be used if you have an active ulcer. If you're receiving anticoagulants, be very careful about taking either aspirin or NSAIDs since both increase the effect of the anticoagulant, and leave you vulnerable to bleeding.

Here's the bottom line. The NSAIDs are my first choice for the relief of joint aches, mild pains, simple headache, and menstrual cramps. Among the various products in this category, I prefer ibuprofen (Motrin, Advil, Nuprin, Medeprin). The brand names listed above differ only in their strength. Motrin comes in 400, 600, and 800 mg tablets and requires a prescription, while Advil, Nuprin, and Medeprin contain only 200 mg, and can be purchased over the counter. Doctors prescribe as much as 2,400 mg of ibuprofen a day to control pain, but if you need more than 800 mg, I suggest you try a different NSAID rather than raising the dosage of the ibuprofen. Sometimes one preparation works when another doesn't. When the over-the-counter 200 mg strength of ibuprofen 3 or 4 times a day does not afford enough relief, I recommend the following that do require a prescription—Naprosyn (naproxen), 500 mg, twice a day, or Voltaren (diclofenac), 50 mg, 3 times a day, both after meals. The latter is probably the most widely used agent in this category world-wide.

If you suffer from chronic arthritis, should you use the NSAIDs around the clock or only when you hurt? I suggest you take them only when necessary. Even though in certain patients with rheumatoid arthritis the NSAIDs *may* possibly slow down the disease process, that has not really been proven as yet, and until it is, it seems to me that the risk of their continued use outweighs the theoretical benefit.

Most NSAIDs begin to work in about an hour, and their effect lasts for some 6 hours, so you'll usually need several a day. (The exceptions are Feldene (piroxicam), a long-acting preparation taken only every 24 hours, and Dolobid (diflunisal), taken every 12 hours.)

Any NSAID can give you a skin rash and other allergic reactions. Some get into mother's milk (Naprosyn), others do not (Motrin). So if you're breast feeding, double-check the NSAID you're using with your doctor. And if you're pregnant, although these drugs have not, to my knowledge, ever been shown to result in fetal abnormalities, I'd avoid them.

The Narcotics

The specific narcotic you receive for pain is pretty much up to your doctor since there's no way you can get any of them legally without a prescription. Narcotics are either mild—codeine, propoxyphene (Darvon)—or strong—morphine, meperidine (Demerol), hydromorphone (Dilaudid), methadone. If you take enough for several weeks or more, any one of them can cause physical and psychological dependence. But that complication has, in my opinion, been overemphasized, and as a result, doctors are unduly hesitant about giving enough narcotic analgesia when it's truly needed over the short term or for terminal cancer patients. Personally, I have no qualms about making a narcotic available in whatever dosage is necessary to anyone with an incurable malignancy.

Morphine is the most effective narcotic, but be very careful about asking for it if you have lung, neurological, or intestinal problems. It can depress your breathing, affect the level of consciousness, and dramatically constipate you. In chronic severe pain of cancer, I recommend MS Contin, a slowly released morphine capsule that is administered every 8 or 12 hours, thus eliminating the need to call for pain killers every 3 or 4 hours.

For pain that doesn't require morphine, but is more severe than what aspirin, Tylenol, and the NSAIDs can control, *codeine* is the best narcotic. It's much less addicting than morphine or Demerol, and is particularly effective for certain kinds of chest pain—like pleurisy (the sharp stitch you feel when you take a deep breath in) or when you've broken a rib. I rarely prescribe propoxyphene (Darvon), because I have found it less effective and more habituating than codeine.

In order to get by with a smaller dosage of a narcotic, and to lessen your apprehension about pain, take an antihistamine such as Phenergan (promethazine) in a 50 mg dose, or 25 mg of Vistaril (hydroxyzine) along with it.

Finally, it has been my experience that anyone who "must have" narcotics for longer than 4 to 6 weeks—and *doesn't* have cancer—is usually addicted.

INDEX

About the Author

ISADORE ROSENFELD, M.D., is an attending physician at The New York Hospital and Memorial Sloan-Kettering Cancer Center and a Clinical Professor of Medicine at Cornell University Medical College. He maintains a consulting practice in Manhattan. Dr. Rosenfeld has for years been a health adviser to millions of Americans who have been counseled by him on national television. He has served as a consultant to the National Institutes of Health on task forces dealing with arteriosclerosis, sudden cardic death, special devices, and hypertension and lipids, and was recently appointed by the Secretary of Health and Human Services to the Practicing Physicians' Advisory Council. In addition to his best-selling books expressly written for patients—*The Complete Medical Exam, Second Opinion, Modern Prevention* and *Symptoms*—Dr. Rosenfeld has written many scientific papers and coauthored a textbook on cardiology. He is an Overseer at Cornell University Medical College, a Visitor at the University of California Medical School at Davis, and Past President of the New York County Medical Society. Dr. Rosenfeld and his wife, Camilla, live in New York City and Westchester County, New York, where their four grown children and spouses visit frequently.